PENGUI

THE HUMAN NA

Theodore Xenophon Barber, Ph.D., is a research psychologist noted for his pioneering experimental investigations into the mind-body
problem and other aspects of consciousness. He
has published almost two hundred articles in
scholarly journals, and is the author of six
books. His research has been conducted at Harvard University, the Worcester Foundation for
Experimental Biology, the Massachusetts De-
partments of Mental Health and Public Health,
and the Research Institute for Interdisciplinary
Science, of which he is currently the Director.

The
HUMAN
NATURE
of BIRDS

A Scientific Discovery with
Startling Implications

Theodore Xenophon Barber, Ph.D.

PENGUIN BOOKS

PENGUIN BOOKS
Published by the Penguin Group
Penguin Books USA Inc., 375 Hudson Street,
New York, New York 10014, U.S.A.
Penguin Books Ltd, 27 Wrights Lane,
London W8 5TZ, England
Penguin Books Australia Ltd,
Ringwood, Victoria, Australia
Penguin Books Canada Ltd, 10 Alcorn Avenue,
Toronto, Ontario, Canada M4V 3B2
Penguin Books (N.Z.) Ltd, 182–190 Wairau Road,
Auckland 10, New Zealand

Penguin Books Ltd, Registered Offices:
Harmondsworth, Middlesex, England

First published in the United States of America by St. Martin's Press 1993
Reprinted by arrangement with St. Martin's Press
Published in Penguin Books 1994

1   3   5   7   9   10   8   6   4   2

THE LIBRARY OF CONGRESS HAS CATALOGUED THE HARDCOVER AS
FOLLOWS:
Barber, Theodore Xenophon.
The human nature of birds/Theodore Xenophon Barber.
p.   cm.
ISBN 0-312-09308-X (hc.)
ISBN 0 14 02.3494 2 (pbk.)
1. Birds—Behavior.   2. Anthropomorphism.   I. Title.
QL698.3.B35   1993
598.251—dc20   93–681

Printed in the United States of America
Set in Cochin
Designed by Jaye Zimet

To the memory of my four wonderful, illiterate Greek grandparents who showed me long ago that there is an intelligence that is deeper than words

CONTENTS

Acknowledgments   *xi*
Introduction   *1*

1. AVIAN INTELLIGENCE ◆ 3

Alex, the intelligent parrot — Avian conceptual abilities —
Exceptional avian memory — Birds using tools — Avian altruism

2. AVIAN FLEXIBILITY ◆ 14

Flexible territoriality — Adaptable foraging — Flexibility in win-
ter and summer homes — Resiliency in nesting and nest repair —
Flexible protective strategies — Intelligently teaching the young —
Resilient mate choice — Intelligently limiting offspring

3. INSTINCTS GUIDE BOTH BIRDS AND HUMANS ◆ 23

Instinctual background of human language — Instincts guide the
human newborn — Instincts of the cuckoo chick — Intelligent im-
plementation of instincts

4. AVIAN LANGUAGES ◆ 33

Human body language — Avian body language — Avian call
language — Avian song language — Avian conversations

5. LORENZO, THE COMMUNICATIVE JAY ◆ 40

6. AVIAN MUSIC, CRAFTS, AND PLAY ◆ 46

*Avian music—Avian aesthetic sense—Avian craftsmanship—
Avian fun, play, and dance*

7. AVIAN NAVIGATION ◆ 58

*Using the sun as a compass—"Reading" the stars—"Reading"
the wind and the weather—"Reading" visual landmarks—
"Reading" the earth's magnetism—"Reading" odors, infrasounds, and other subtle cues—Sensibly deciding to migrate—
Learning with experience—Overview of avian navigational
abilities—Human navigational abilities—The instinctual background*

8. PERSONAL FRIENDSHIPS BETWEEN HUMANS AND BIRDS ◆ 74

*A talking starling—A very sociable jackdaw—Friendship be-
tween a professor and an owl—Three parakeets with "human"
personalities—Close friendships with many wild birds*

9. OVERVIEW OF BIRD INTELLIGENCE ◆ 98

Avian superiorities—Human superiorities

10. WHY BIRDS HAVE BEEN TOTALLY MISUNDERSTOOD ◆ 104

*Ignorance of birds as individuals—The fallacy of size—The
small-brain fallacy—Misunderstanding of instincts—
Misunderstanding intelligence—Human narcissism—Human
benefit—Anthropomorphic dread*

11. ARE ALL ANIMALS INTELLIGENT? ◆ 118

*Intelligent apes—Intelligent cetaceans—Intelligent fish—
Intelligent hymenoptera*

12. REVOLUTIONARY IMPLICATIONS OF ANIMAL INTELLIGENCE ◆ 149

Humanity, the destroyer—Humanity renewed

APPENDIX A. THE CONTINUING COGNITIVE
ETHOLOGY REVOLUTION ◆ 161

*A clarification for my colleagues in the behavioral and brain
sciences*

APPENDIX B. HOW YOU CAN PERSONALLY EXPERIENCE A BIRD AS AN
INTELLIGENT INDIVIDUAL ◆ 165

*Befriending wild birds — Befriending birds living freely in your
home — The importance of befriending birds*

APPENDIX C. SCIENTIFIC NAMES OF SPECIES ◆ 173

Notes   *177*

Index   *219*

ACKNOWLEDGMENTS

In writing this book I owe a profound debt to the work of many pioneering investigators. As detailed in Appendix A, the book would not have been written without the stimulating writings of Donald R. Griffin, the founder of cognitive ethology. Also, the data that give the book its substance (but not the interpretations, for which I alone am responsible) derive primarily from the splendid research and scholarship of sixty-five investigators to whom I am deeply indebted: Kenneth P. Able; Thomas Alerstam; L. R. Aronson; R. Robin Baker; Russell P. Balda; Luis Baptista; Colin G. Beer; Thomas Berry; Peter Berthold; Lester R. Brown; Rémy Chauvin; Noam Chomsky; Nicholas E. Collias; William R. Corliss; Frans de Waal; Paul R. Ehrlich; Irenaus Eibl-Eibesfeldt; Peter D. Eimas; Stephen T. Emlen; Roger S. Fouts; Matthew Fox; Beatrix T. Gardner; Howard Gardner; R. Allen Gardner; Frank B. Gill; Jane Goodall; James L. Gould; Charles Hartshorne; Bernd Heinrich; Louis M. Herman; Richard J. Herrnstein; Bert Hölldobler; Len Howard; Alan C. Kamil; Melvin L. Kreithen; Thomas S. Kuhn; Robert Franklin Leslie; Martin Lindauer; Hubert Markl; Peter Marler; Andrew N. Meltzoff; Emil Menzel; Fernando Nottebohm; F. Papi; Francine G. Patterson; Roger Payne; Irene M. Pepperberg; David Premack; Karen Pryor; Jeremy Rifkin; Carolyn A. Ristau; Bernard E. Rollin; Kirkpatrick Sale; E. Sue Savage-Rumbaugh; Klaus Schmidt-Koenig; Thomas D. Seeley; Alexander F. Skutch; David Suzuki; Herbert S. Terrace; Brian Tokar; Charles Walcott; Joel Carl Welty; Edward O. Wilson; Sheryl C. Wilson; and W. Wiltschko.

I thank Lee Carlton, Thomas J. McCormack, Carol T. Viera, Sheryl C. Wilson, and Morton Yanow for critically reading the manuscript.

The
HUMAN
NATURE
of BIRDS

INTRODUCTION

After a thirty-year career as a behavioral scientist, with a professional reputation as a hard-headed skeptical researcher, I turned my full attention to the specialized topic of animal intelligence. During six years, as I studied the accumulated scientific literature, I gradually realized that, in numerous research projects, birds had behaved with intelligence, purposiveness, and flexibility. Since I had previously accepted the official scientific view that birds are instinctual automata, I was horrified to realize that I and virtually all other scientists have been blocked by the official taboo against anthropomorphism from perceiving the nature of reality, beginning with the intelligent nature of our close neighbors, the birds. The research data clearly showed what official science vehemently denied: Birds are sensitively aware and emotional; they have distinctly different personalities; and they know what they are doing.

In this book I first summarize the research with birds that challenges the dominant world view which excludes consciousness and intelligence from nature. Then I turn to the scientific research demonstrating unexpected intelligent awareness in other animals. In the final chapter I discuss the revolutionary implications of animal intelligence for our science and philosophy, for our individual lives, and for the future of our civilization.

CHAPTER ONE

Avian Intelligence

As I analyzed and synthesized the research data on the migration of birds, their learning capabilities, and their characteristic behaviors, I realized that the scientific findings converged on three unexpected conclusions:

1. Birds have many abilities that humans assume are unique to humans, including musical ability (appreciation, composition, and performance),[1] ability to form abstract concepts, ability to use intelligence flexibly to cope with constantly changing life demands,[2] and ability to play with joy and mate erotically.

2. Although humans are superior to birds in certain kinds of intelligence (such as symbolic-linguistic intelligence), birds are superior to humans in other kinds of intelligence (such as navigational intelligence).

3. Birds are not only intelligent, aware, and willful; they also can communicate meaningfully with humans and relate to them as close, caring friends.

These revolutionary conclusions will be thoroughly documented in Chapters 1 to 10. First let's look carefully at five sets of recent research projects that have revealed the true nature of birds.

Alex, the Intelligent Parrot
◆

For the past fifteen years, Alex, a sixteen-year-old African gray parrot,[3] has been living in Professor Irene M. Pepperberg's research laboratory, where he has been carefully observed and studied.[4] Alex uses English words to communicate in ways previously thought to be the sole domain of humans. He *understands* what the experimenters (faculty and students) say to him, and he speaks to them *meaningfully*. As we shall see in a moment, he uses meaningful and, at times, creative English words and phrases to request what he wants, ask sensible questions, answer abstract questions, tell experimenters to go away and leave him alone, and in many other ways relate with people in a way that is expected of a person but not of a bird.

Alex is not isolated in the laboratory, and he exercises much control over his life. He is out of his cage about eight hours each working day. During this time he interacts with the experimenters, who treat him respectfully, as an important member of the laboratory team, while teaching him words and concepts and testing him in formal, rigorously controlled experiments. The experimenters also respond to Alex's requests for particular foods or toys (for example, "Want cherry" or "Want truck") and they answer his requests for particular information ("What color?"). Because his wings are clipped, he is unable to fly, so he asks to be taken to specific places ("Wanna go gym"). The experimenters take him where he wants to go.

Alex learned to say meaningful things in English primarily by observing two human models speaking to each other about objects and their characteristics. He listened much of the time as two experimenters near him took turns methodically questioning each other. One experimenter asked and the other answered such questions as "What shape is the wood?" "How many?" and "What object is blue?" The experimenter asking the question commended the other for correct answers ("yes," "good") and expressed disapproval for incorrect answers. To emphasize correct pronunciation, an experimenter at times intentionally mispronounced a word and was immediately given "corrective feedback" by the other experimenter while Alex observed the interaction. Alex participated in these interactions on his own initiative, at times answering the question that was asked or, less often, asking his own question. He received praise for his correct answers and was given the object he named correctly as his reward unless he refused it and vocally requested another object he desired as a reward substitute.

In answering difficult questions, Alex utilizes his surprisingly large vocabulary, which includes the names of more than one hundred things in his environment. For example, he says in the appropriate context the words for water, grape, pasta, chair, grain, shower, block, box, shoulder, knee, key, chain, tray, gravel, plus eighty-six or so other things. He also correctly uses numbers up to six, and he knows the words for the significant characteristics of test objects, such as their colors (green, red, blue, yellow, gray, orange, and purple), their shape in terms of corners (two-, three-, four-, five-, or six-cornered), and the materials they are comprised of (cork, wood, paper, chalk, wool, rock, and rawhide).

He makes personal requests such as "You tickle me" or "I want popcorn." In fact, he requests any one of his fifteen foods — "I want some tea," "I want banana." Although he is provided with water and basic parrot seeds, he receives fresh fruits, drinks, vegetables, and special nuts (cashews, almonds, pecans, walnuts) only if he asks for them. (Even though they are all stored out of sight, he remembers them perfectly well and often asks for them.) If given a food other than the one he requests, he refuses it and repeats his initial request.

He asks to be taken to particular places — "Wanna go chair," "Wanna go gym" — and someone takes him. If taken to the wrong place, he refuses to get off the transporting arm and repeats his request. He may ask ("I want . . .") for any one of his objects or toys. He requests his "peg wood" to chew on, a "key" for scratching himself, or a "cork" to clean his beak. If the experimenters offer him something other than what he asks for, he says "No" and repeats his request. Alex exerts control over his life by saying "No!" to refuse anything he does not want. He also tells the experimenters when he wants them to "Come here." He requests particular information: "What's this?" "What shape?" "What's here?" "You tell me. What color?" He asks students what a certain thing is called and what material it is made of.

In a typical experiment, Alex is asked "What object is gray?" and is shown a tray containing seven items (for example, a purple key, yellow wood, green hide, blue paper, orange peg wood, gray box, and red truck). Alex looks carefully at the seven items and answers correctly "box." He is then asked "What same?" as he is shown the tray now containing a red paper triangle and a blue wooden triangle. Alex again studies the objects and answers correctly "Shape." Then Alex is shown three corks and two keys intermingled on the tray and is asked, "How many cork?" He correctly answers "Three." Over a series of experiments he is shown various objects in many combinations and answers correctly an astonishing number of questions: What object is orange? . . . rose? . . . green? . . . yellow? . . . purple? . . . blue? What color is key? . . . peg wood? . . . truck? . . . block? . . . wool? . . . box? . . . cup? . . . wood? . . . chain? . . . hide? . . . chalk? What object is two-corner? . . . three-corner? . . . four-corner? . . . five-corner? . . . six-corner? What shape [number of corners] is paper? . . . wood? . . . hide? . . . rock? . . . wool? . . . key? . . . plastic? Since Alex never knows what he will be asked next, he must pay close attention to each question, understand exactly what is being asked, and be ready to answer questions about anything he has ever learned. (For example, a question of "What's different?" might be preceded by "What color?" and "What's this?" and followed by "How many?" "What shape?" and so on.)

Alex spontaneously uses his knowledge of English to construct simple sentences and to create his own meaningful names for new objects. He creatively combines words that he knows into new sensible phrases, such as "banana cracker" for banana chip, "gray nut" for gray sunflower seed, "rock nut" for the rocklike Brazil nut in its shell, "cork nut" for the corklike almond in its shell, "banerry" for an apple that tastes somewhat like a banana and looks like a very large cherry, and "rock corn" for dried corn (which he distinguishes from fresh corn, which he calls "corn"). After he had learned the numbers up to six and had learned to call a triangle "three-corner" and a square "four-corner," he creatively called a football "two-corner," a pentagon "five-corner," and a long rectangular piece of computer paper "four-corner." He asks questions to obtain information. For instance, he learned the color "gray" when he asked a student to tell him the color of his reflection in a mirror. He simultaneously learned the color "orange" and the name "carrot" when he asked a student eating a carrot what color it was and what it was called. Henceforth he reliably identified "gray" and "orange," and he added carrots to the list of vegetables he requests from time to time.

Alex is not passive; to a large extent he actively controls his life. For instance, when a desired nut was placed by an experimenter under a heavy metal cup that Alex could not lift, he told the experimenter "Go pick up cup." As soon as the experimenter picked up the cup, Alex promptly walked over and ate the nut. If the experimenters fail to make the experimental tasks sufficiently interesting for him, Alex lets them know he is bored. He communicates his boredom by asking to be moved to a new location or by ignoring the experimenters and preening. During an especially long and difficult experimental session, Alex communicates his frustration and desire to stop answering questions by requesting to return to his cage: "Wanna go back." When brought to his cage, he tells the experimenter "Go away" and climbs inside the cage onto the swing, ignoring the experimenter and refusing to interact with anyone.

Alex understands what he is saying, and, when he answers

a question correctly, he demonstrates confidence in his answer. About 5 percent of the time, student experimenters have mistakenly scolded him for a correct response. Not accepting their reprimands, Alex confidently repeated his *correct* answer.

In brief, Alex the parrot acts surprisingly like a human when he speaks to obtain what he desires, to control his environment, to resist what he does not want, to ask questions, and to learn about things he does not know. He is far more intelligent, far more like a human, than anyone had imagined a bird could be.[5]

Avian Conceptual Abilities
◆

The surprising conceptual ability shown by Professor Pepperberg's parrot has also been demonstrated in pigeons. Investigations initiated by Professor Richard J. Herrnstein at the Harvard Psychological Laboratories yielded surprising results.[6] Laboratory pigeons conceptualize at a high level of abstraction. When given food only when they peck at projected photographic slides containing a human, pigeons show that they include under the concept of "human" both male and female humans of every race, culture, color, chronological age, and size. They also identify the back of a human head, a human hand or foot, and other parts of the human anatomy as "human." When given food for recognizing a particular human, the pigeons recognize the human under various "disguises," for instance, when the person is nude, or wears strange clothes, or is a tiny face in a large group photograph.

Each pigeon concept that was assessed in the laboratory was surprisingly as general and complete as typical concepts of humans. When fed when they pecked at "water," pigeons pecked at photographic slides of an ocean, a brook, tap water, and a tiny puddle on a muddy road (and they did not peck at a slide of the same muddy road without the puddle). The concepts of pigeons are not only highly differentiated but also hierarchical; pigeons showed by pecking that they have concepts for oak tree, maple tree, tree trunk, tree branch, tree leaf, etc. Laboratory pigeons also distinguished man-made objects (such as streets and buildings) from natural objects (such as forests and fields), and they even distinguished esoteric objects, such as equilateral triangles from other kinds of triangles.[7]

Laboratory pigeons also succeed at other tasks that were thought to be unique to humans or to primates. Pigeons could "hold in mind" a particular sequential order, say BDAC or CBAD, and then peck at sets of four objects in that particular sequence to receive food. They also behave as if they possess the mental faculty of insight when they put together a series of known elements in creative ways to solve a new problem.[8] They also perform well (at least as well as mammals) on complex tests that psychologists view as tests of intelligence. These tests include: *multiple reversal tests* (requiring the flexibility to switch back and forth intelligently between something that is now "right" and rewarded but was just recently "wrong" and unrewarded); *learning set tests* (requiring the ability to learn a general principle for solving different problems); and *oddity tests* (requiring the ability to pick the odd item in each set of three objects).[9]

Laboratory pigeons learned to recognize each of the twenty-six letters of the English alphabet. It seemed odd to the researchers that the birds made the same kinds of initial mistakes as elementary school students; for instance, they tended to confuse C and G and also W and V.[10]

Other avian species tested in laboratories (canaries, parrots, jackdaws, woodpeckers, great tits, and chickens) also showed surprising humanlike abilities on formal tests. These abilities included counting at least up to eight, understanding very abstract concepts such as "symmetry" and "uniqueness," communicating a desire for any one of five foods by drumming differently for each kind, and handling complex problems that have more than one factor varying at the same time.[11]

In brief, the prevalent notion that concepts are unique to humans is not valid. Laboratory birds have conceptualized what they were "requested" to conceptualize. It appears likely that birds in nature conceptualize everything that matters to them.

Exceptional Avian Memory

◆

Another cognitive function—the ability to remember—has also attained a surprisingly high level in birds. They remember not only the exact locations of their summer and winter homes on different continents, but also the exact location of the many items they store for winter use and the exact location of particular flowers they previously emptied of nectar. In fact, some species surpass humans in long-term memory. To survive the coming cold and heavy snow of winter, each individual Clark nutcracker buries thousands of seeds in separate caches during autumn and then remembers the precise location of each cache many months later in winter. These birds remember large numbers of exact locations (more proficiently than humans) by noting, at the time of seed burial, the precise distance and direction of the surrounding small and large objects, retaining the many bits of data, and recalling each bit much later.[12]

Birds Using Tools

◆

The unique human hand is wonderful for manipulating tools; in fact, humans are tool-making and tool-using specialists. However, the common assumption that only humans have the intelligence to use tools is false. Birds also can make tools or use selected objects as tools to attain a goal. For example, each male satin bowerbird colorfully paints the walls of his bower after he finds some kind of fibrous material that can be used as a brush and some kind of color-producing substance (such as cherries or charcoal) that can be used as paint.[13] Woodpecker finches select a twig (about an inch or two long), straighten it by breaking off tiny pieces, hold the twig in the beak, and poke it into cracks and scrape it around crannies until an insect is flushed out. Then the bird tucks the twig very quickly between its claws and devours the insect.[14] Green jays, North American nuthatches, and various other birds have been seen to catch insects in a similar way using miniature tools they construct from thin pieces of wood, thorns, or cactus spines.[15]

Corvids (crows, jays, jackdaws, ravens, rooks) are especially likely to make useful tools from selected materials. They have been reliably observed selecting available sticks or strips of newspaper to make a rake for pulling in grain from outside the cage; placing solid objects in their drinking dish to raise the level of water to where they can drink it; making a water bath for themselves by inserting a plug of the right size in an appropriate plug-hole in the aviary; and using a plastic cup to fetch and pour water on their food when it is too dry.[16] Also, jays have been reliably observed to crack walnuts by placing each nut in an exactly fitting holdfast (formed by two cross branches), using one foot to keep the nut steady, and lining up the nut so that its seam splits when a few sharp blows are delivered with the beak.[17] Ravens have defended their nests by picking up stones in their bills and tossing them one after another at human intruders.[18]

When crows drop mollusks on rocks from up high, they maximize the possibility that the shell will break by simultaneously taking three variables into consideration—the size of the mollusk shell, the distance to the ground, and the characteristics of the rock. Similarly, crows at times drop palm nuts on highways, then wait for passing cars to crack them.[19] In Scandinavia hooded crows at times catch fish by using the lines that fishermen leave suspended through holes in the ice of frozen lakes. The bird seizes the line in its beak, walks off with it away from the hole, then walks on top of the line back to the hole (thus preventing it from slipping), and repeats the process carefully until the fish is brought up.[20] Other kinds of birds fish with bait; for instance, herons at times use insects or bread bits as bait and stab the fish when it approaches the bait. A recent elegant study showed that the green-backed herons that fished with bait differed in their preferred methods; they used different kinds of bait and handled it in different ways.[21] An esteemed senior scientist, Professor Donald R. Griffin of Rockefeller University, has dared to state that when a heron is bait-fishing in its own unique way, it may think about what it is doing and may even look ahead to what it hopes to achieve.[22]

Egyptian vultures and Australian black-breasted buzzards have been reported to use a creative variety of methods to break open hard-shelled eggs they wish to eat. They may either hammer the egg with a stone held in their beaks, drop stones on it as they fly over it, throw stones at it from the ground, or toss it a considerable distance.[23] Blackbirds, thrushes, bee eaters, and eagles have also been reliably observed using stones intelligently to stun their prey, break shells, and drive away intruders. To uncover food, at least two finch species on the Galapagos Islands at times push aside seemingly immovable large stones with their feet after bracing their heads against a large rock for leverage. The black kites of India are called "fire hawks" because they often are seen picking up smoldering sticks from fires, dropping them on dry grass, and then waiting to catch small animals as they run out of the grass to avoid the fire.[24]

An important conclusion emerges from these data: Birds are capable of *intelligently* using tools or tool substitutes — selecting and using external objects "as extensions of themselves."

Avian Altruism

◆

People assume that only humans are capable of such commendable qualities as caring for others, empathy, and altruism. This assumption is mistaken. There are many documented cases of birds acting in a caring or empathic way. Reliable witnesses have observed numerous events such as the following.

Pairs of terns took turns in relays lifting a disabled flock-mate, each holding on to one of its wings, and carrying it to safety.

A male European robin fed and kept alive a rival male bird after crippling him in a fight.

A male jay guided a human over a considerable distance to an abandoned newborn of a different avian species that had fallen from its nest. The jay then helped the human raise the young bird to maturity by protecting it and sharing its own precious food with it.

Crippled crows called frantically when hunters were ready to club them to death; their cries were answered by flockmates who attacked the hunters and forced their retreat. Other avian species have also been reliably observed to act contrary to the self-preservation instinct when flockmates were wounded, staying nearby and uttering distress cries even though they were in danger of being shot.[25]

Many other behaviors that are common in birds would be interpreted as caring, concern, and love for others if seen in humans. Separated lovebirds act as if they are pining for an absent mate and appear very joyful when unexpectedly reunited. In many avian species unmated birds go to active nests and help the parent birds feed their voracious nestlings.[26] Other instances of avian caring, empathy, and altruism will be presented later in this book.

◆

CHAPTER TWO

Avian Flexibility

The common assumption that the behavior of birds is "mechanical, stereotyped, and instinctive"[1] has been shown to be invalid. The truth, which is much more complex, has two complementary dimensions:

◆ Each bird, in the same way as each human, inherits the essential anatomical and behavioral plans or programs of its ancestors.
◆ Each bird and each human implements its inherited, instinctual plans or predispositions in its own constantly changing life situation in which it has to act flexibly (intelligently) to survive.

The instincts of birds and humans will be discussed in the next chapter. In the present chapter we will look at avian flexibility in critical life activities, such as defending and monopolizing a territorial area that has a sufficient food supply, constructing a secure nest, and protecting and guiding the young.

Flexible Territoriality

◆

A recent acclaimed text summarizes the data on territoriality in this way: "Ornithologists thought that the territorial behavior of birds was genetically programmed and static. In fact, territorial behavior is flexible and dynamic."[2] Birds that have been carefully observed (storks, blackbirds, golden-winged sunbirds, pomarine jaegers) vary their territorial behavior flexibly and intelligently: The size of the area defended and monopolized by each bird increases as the food supply decreases; newcomers are driven off when food is scarce but not when it is abundant; the actions of the territory-holder vary appropriately depending on the intentions of the intruder; and the supposedly rigid territorial instincts totally disappear when food is either very abundant or extremely scarce.[3] In brief, birds intelligently or flexibly maintain sufficient territory to meet their needs.

Adaptable Foraging

◆

Humans living in hunting and gathering societies intelligently and flexibly adjust their search for food to exploit new food sources as they arise. Birds also adjust their food search in essentially the same way. Great tits in the laboratory searched deliberately and intelligently for concealed food and flexibly shifted their foraging strategies rapidly with quickly changing circumstances.[4] The foraging strategies of gulls are intrinsically creative since they find and exploit many new food sources — thronging city dumps, flying to the one particular place where great numbers of earthworms rise to the surface after a heavy rain, following refuse scows with food scraps out to sea, and so on.[5]

Birds can flexibly shift their age-old food preferences when it is the intelligent thing to do. For instance, kea parrots in New Zealand shifted from a diet of fruits and seeds to a lamb diet after the settlers began raising sheep.[6] When tin foil and cardboard tops were first introduced to cover milk bottles in Britain, at least eleven species of tits quickly learned how to remove the covers. They began following the delivery vehicles on their routes and made milk a major part of their diet.[7]

Parent birds intelligently adjust their food-gathering trips to meet the needs of their growing young. In fact, by flexibly varying their selection of foods, the parent birds in a recent investigation were able to keep the number of meals brought to each nestling practically constant (7.1 to 7.5 meals per hour per nestling) even though the number of nestlings was experimentally varied and the amount of food each consumed changed over time. Very small items were selected for the newly hatched nestlings, and the size of the items was flexibly increased gradually as the young grew.[8]

Flexibility in Winter and Summer Homes

◆

The common viewpoint that sees birds as machinelike creatures governed by inflexible instincts is also contradicted by the recent discovery that birds typically behave differently in their "winter" and "summer" homes.[9] Birds that are aggressive, or territorial, or social, or insect eaters in their northern homes often drastically change their behaviors—becoming nonaggressive, or nonterritorial, or nonsocial, or fruit eaters—when they migrate to the tropics. In its northern home, the Eastern kingbird is a feisty insect eater that pugnaciously drives away other birds from its territory; when it migrates to South America in autumn, it becomes nonterritorial and nonaggressive and lives socially in a flock that feeds on a particular fruit. In North America, broadwinged hawks are active during the day eating mice, small reptiles, and small birds; after they migrate to Panama, they are active at twilight, feeding upon a particular insect. The Cape May warbler eats insects in the North but, when it winters in the

Caribbean, it flexibly switches to nectar and juices sucked from fruit.

The flexible behavior of birds is evident in the tropics where they may change their lifestyles several times as conditions change. During the rainy season bay-breasted warblers feed alone on insects in young forests; at the beginning of the dry season they combine into flocks in mature forests (where greater retention of moisture allows more insects to survive); at the end of the dry season they return to the young forests and in flocks exploit the abundance of fruit. Each shift in behavior is sensible; for instance, when food is plentiful, clustering in flocks is an intelligent antipredator strategy.

Resiliency in Nesting and Nest Repair
◆

Many kinds of birds have been reliably seen to adjust flexibly to new circumstances by radically changing where they build their nests. Species that typically nest in trees alter their supposedly rigid nest-building instincts in treeless countries and build their nests on the ground or among rocks.[10] The peregrine falcon preferentially nests in cliffs, but it nests in trees when cliffs are unavailable and on the ground when both cliffs and trees are lacking. Similarly, depending on circumstances, the heron may build its nest on a tree or cliff or in a marsh.[11] When new predators were introduced in Samoa, tooth-billed pigeons that had constructed their nests on the ground intelligently shifted to building them in trees.[12]

Although sparrows are widely distributed around the world, only those living in certain parts of Africa surround their nests with thorns to protect against apes, reptiles, and other predators. In fact, there are so many tree-dwelling predators in the tropics that more than half of the bird nests there are covered on top and have a side or bottom entrance that make access by predators difficult.[13]

Various species maximize their security by constantly

changing the locations of their nests.[14] Pigmy nuthatches, which roost together during the winter in cavities excavated by woodpeckers, choose a different cavity each night by intelligently taking into account the temperature, the precipitation, and the size of the group.[15] Black-headed weaver birds protect their nests from predators by constructing them in exceptionally safe locations, such as next to a hornet's nest, on tree limbs overhanging streams, and on human habitations. Many different kinds of birds—tropical woodpeckers, black-throated warblers, ruddy kingfishers, caciques—scare off predators by building their nests next to wasp or hornet nests or inside the nests of biting ants or stinging bees.[16]

Birds behave sensibly when their nests are damaged. A typical observation is a pair of house martins repairing their seriously damaged nest by appropriately timing and coordinating their trips to bring mud for the nest and insects for the young. When experimenters have deliberately damaged bird nests, the birds did just what was necessary to repair the nest—no more and no less. After experimenters removed the soft inner lining material from the nests of birds brought to the laboratory, the birds ingeniously replaced it with soft feathers pulled from their own bodies. When experimenters supplied the kind of material birds ordinarily gather for their nests, the birds intelligently used that material instead of robotically continuing to gather it themselves. As Donald R. Griffin has forcefully emphasized, birds do not follow rigid or stereotyped sequences when their nests are damaged; instead, they act flexibly, sensibly to repair the damage.[17]

Flexible Protective Strategies

◆

Many birds have been reliably seen to protect their young by using methods that humans assume are unique to humans. When a flood threatened, a mated pair of nightingales marvelously coordinated their actions moment by moment to fly their nest to a safe place while bearing it between them. Numerous birds of many different species have been seen to use their backs, feet, legs, or beaks to transport their endangered eggs or nestlings to a safer place.[18]

When a snake, hawk, or other predator appears, a crowd of smaller birds often gather nearby, and each bird protests individually by repeating its own alarm cry while it flits from twig to twig and threatens the intruder by darting close to it or dive-bombing it. Sometimes the predator is attacked by all the birds en masse. These mobbing activities increase dramatically after the eggs are laid and are clearly aimed at protecting the young. Young birds learn how and whom to mob by observing their elders.[19]

Careful observers have documented that parent birds utilize a variety of sensible strategies to divert a predator's attention to themselves and away from helpless offspring.[20] The parent may go to an observable spot away from the nest and then act as if it is sitting on a nest; it may act like a small rodent scurrying away in a zigzag course (but glancing back from time to time to see if the predator is still in pursuit); it may walk away from the nest while chirping excitedly; it may pretend it is unable to move by beating its wings helplessly; or it may act as if it has been injured by dragging one wing near the ground. The predator often follows the seemingly injured bird. When it has been led sufficiently far away from the nest, the bird suddenly flies away. When using the injury-simulation strategy, birds monitor the intruder's movements and its eye gaze (toward or away from the nest) and modify their behavior accordingly. They also always move in the sensible direction away from the nest.[21]

These predator-diversion strategies are not rigid. Individual birds prefer particular strategies and flexibly vary their strategies. For instance, of two neighboring birds of the same species, one will choose to use the injury-simulation strategy with every human intruder while the other bird may never use it. The same bird that always uses this strategy with humans will flexibly defend the nest in a totally different but very effective way when the intruder is a snake (attacking it directly), when it is a flying bird of prey (fleeing from it), and when it is a horse or cow (laying low until the last moment and then flying suddenly into its face with a great outcry). Respected senior investigators, such as Alexander Skutch and D. R. Griffin, who have studied the data carefully, agree that the parent bird is "cognizant of exactly what it is doing" when it carries out a strategy to prevent harm to the young.[22]

People are always shocked when they first realize that birds are not only aware of them but also observe them carefully. In localities where they are persecuted by humans, jays and crows have guards or sentinels that scrutinize each human. If they see a person with a gun, they leave the area immediately, and they definitely recognize the person again without the gun. They make fine judgments as to the potentially harmful intentions of people; for instance, they distinguish between a person holding a gun and one holding a stick.[23]

Birds of the same species react very differently toward people depending on how people act toward them. Pigeons are tame in cities where they are fed and protected, but they flee from people in the nonurban areas where they are hunted. When humans first approached birds on isolated islands, the birds allowed them to approach closely.[24] A very important reason why birds in nature have long hidden their intelligence from almost all humans is because they have learned to fear humans and flee if they are approached too closely.

Intelligently Teaching the Young

◆

Young humans and young birds typically learn from their parents what to fear, whom to regard as an enemy, what to eat and drink, and where and when to find food and other necessities.[25] Some aspects of this learning can be described as "formal education." Parent birds have been reliably observed teaching their youngsters to avoid the danger of flying into windowpanes by enticing them to peck at the glass and perceive its impenetrability. Parents coax or drive young penguins into the water the first time and give them expert demonstrations of swimming maneuvers. Peregrine falcon mates have trained their young to catch prey by a formal three-step procedure. First, one parent drops the prey it is carrying as it flies above the youngster. Next, the second parent, who flies lower, catches the prey if the youngster's swoop at it fails. The second parent, now holding the prey, switches to fly above the youngster while the first parent now moves below it, and they repeat the process. Other raptors also have been reliably observed to teach their young by dropping prey to them, releasing live small birds for them to catch, and flushing out hidden prey that the young then catch.[26]

Resilient Mate Choice
◆

There are numerous references in the ornithological literature to female birds who either have strong preferences for particular male birds or who show numerous signs of depression, such as listlessness and anorexia, at the loss of a particular mate.[27] In fact, it appears that male and female birds may choose each other in essentially the same way as humans by considering the other bird's personality characteristics and physical attributes. The physical characteristics that are considered in avian mate choice can be delineated rather precisely. For instance, a recent experiment showed that, in choosing a desirable mate, male pinyon jays use specifiable criteria (the female's chronological age, her size, and her past breeding experience) and that the stringency of each criterion is flexibly and intelligently varied with the number of available females.[28]

Intelligently Limiting Offspring
◆

Since many present-day human couples project that their financial resources are insufficient to raise more than one or two children in the style they deem appropriate, they sensibly choose to limit the number of their offspring. A thorough investigation showed that tawny owls make a similar sensible choice each season as to how many young to raise. When food is scarce they do not mate; when the food available is below average, they lay a proportionally smaller number of eggs; and when the food supply is seriously reduced after they lay their eggs, they do not incubate them.[29] Various other species also limit their offspring in a flexible way by delaying courtship, laying fewer eggs, or not feeding the weakest chick in the brood.[30]

At this point the critical reader is bound to ask an important question: Are not the data just summarized, which indicate avian flexibility, contradicted by other data that show that birds act instinctually? In the next chapter I shall summarize the recent revolutionary research that answered this question by demonstrating that both birds and humans are guided by instincts and that the two taxa implement their instinctual predispositions equally flexibly and intelligently.

CHAPTER THREE

Instincts Guide Both Birds and Humans

The common belief that birds act instinctually whereas humans act intelligently has been shown to be incorrect. Research evidence from three disciplines (avian biology, human linguistics, and child development) has converged on a surprising conclusion: Both birds and humans are born with propensities or species' guidelines for behaving like birds or like humans; and both birds and humans intelligently implement their innate species' guidelines or instincts to fit continually varying circumstances.

Instincts are those inborn guidelines or plans of action that guide a bird as it composes a song that follows its species' pattern, or builds a nest that adheres to a generalized species' blueprint, or flies somewhat awkwardly but like a bird on its first attempt, or uses particular species' calls to communicate with its flock, or navigates alone to its ancestral winter home. However, as we will see throughout this book, birds actualize or implement their innate instinctual guidelines by flexibly varying and fine-tuning their behavior in sensible ways to adjust to continually varying circumstances.

Instincts—innate programs or species' plans (transmitted from one generation to the next via the sperm and the egg)—also guide humans. Humans inherit their instinctual predispositions to behave like humans—to walk upright on two feet like all humans; to use hands with the skill characteristic of humans; to understand the rules of human language; to learn to speak like a human; to smile, laugh, cry and experience human emotions; to use human cognitive capabilities; and to behave in many other ways like human beings. As we all know from personal experience, these instinctual programs that guide all humans are fine-tuned and implemented flexibly and intelligently by each one of us to adjust to our particular circumstances.

Although the instinctual background of avian behavior has been profusely discussed in the scientific literature of the past century, only in recent years has scientific evidence converged on the surprising conclusion that human beings are also instinctually guided to carry out the activities that are common to all humans. Since this little-publicized clarification of human instincts provides a needed perspective on the widely publicized instincts of birds, let us look closely at the recent discoveries that revolutionize our understanding of human instincts.

Instinctual Background of Human Language
◆

One of the most astonishing recent scientific discoveries is that the special intelligence of humans, as denoted by their capability for speech and language, is as innate and instinctual as the instinctually guided songs, calls, nest designs, migrations, and mating-parental patterns of birds. The discovery that the uniquely human linguistic intelligence is instinctual is derived from unexpected results of scientific research.

◆ The newborn human is instinctually "pre-tuned" to attend selectively to a small range of sound waves (within a countless number of impinging sound waves) that are peculiar to human speech. In fact, the newborn is innately programmed to pay special attention, or "tune in," to the two dozen consonant sounds used in human speech, including consonants not present in the language it hears.[1]

◆ All human infants, *including congenitally deaf babies with deaf-mute parents*, are instinctually guided to "spontaneously" pass through a cooing stage (making vowel-like, pitch-modulated, squealing, gurgling, cooing sounds). All human infants, including deaf ones, inherit a program to "spontaneously" move on to a babbling stage (making one-syllable utterances, such as ma, da, goo-goo, doo-di-la)! (Later we shall see that the human infant's instinctual babbling program resembles the instinctual babbling program of birds known as "subsong.")[2]

◆ The babbling of the hearing infant, but not that of the deaf infant, gradually changes to match the speech heard in its environment. The infant's ability to articulate the sounds of human speech derives from instinctual guidelines, innate in humans but not in other animals, that *guide* rapid opening and closing of the vocal cords with *coordinated* changes in the pattern of breathing and with *harmonized* contractions of more than fifteen pairs of muscles in the throat, palate, tongue, and lips.

◆ The young human child does not simply imitate the utterances it hears. Instead, it is instinctually guided to learn the "rules" for combining words into grammatical units and thus to utter meaningful sentences it has never heard before. Ironically, this wonderful, creative human capability — to create an endless number of grammatically meaningful sentences (and thus to talk about anything) — is instinctual!

◆ As the leading language scholar Noam Chomsky has repeatedly emphasized, all normal human children learn a human language with apparent ease without the need for special reinforcements or rewards or special tutoring. The only requirement is exposure to human speech. Furthermore, what young children are instinctually guided to learn simply by hearing others speak is tremendously complex. Linguists are in awe of the unending intricacies of the syntax, semantics, phonology, and pragmatics that constitute human language. Neuroscientists are equally awed by the huge number of nerve-muscle interactions that coordinate the formation of each word.

◆ Only humans have nerve cells in special areas of the brain (such as Broca's area, Wernicke's area, and the arcuate fasciculus) that specialize in performing the analyses and syntheses that underlie the comprehension and articulation of speech. The inherited program that guides the formation of these special speech areas specifies that they be localized on one side of the human brain.

The avian communication system, which utilizes calls, songs, and body language, is just as instinctually guided as the communication system of humans.[3] Birds are also set to detect the particular sounds of their species and to pass through a "subsong" period that has important commonalities with the human infants' "babbling" period. Normal birds, but not deaf ones, also match their vocalizations to the innate patterns of their species' song; in some species the matching is immediate, without special practice. Birds also have an instinctual ability, which in most species simply unfolds (without special learning), to understand the communications from the other members of their flock and to communicate to others everything that concerns the flock. Birds also have nerve cells in particular areas on one side of the brain that specialize in song production and song communication.[4]

In brief, both humans and birds have natural, instinctual communication systems that are programmed in much the same way (to cover much the same dimensions).

Instincts Guide the Human Newborn
◆

Recent scientific research on newborn infants and newly hatched birds also has converged on the revolutionary conclusion that humans and birds are guided by instincts in essentially the same way. The instinctual guidelines of the human newborn have been clarified by the following discoveries.[5]

- ◆ If the newborn infant is placed on the mother's stomach for the first feeding, it appears automatically to seek the milk in the breast. Specifically, it tends to push itself forward with its legs, flails its arms, attempts to clutch with its hands, searches with its mouth, and, as soon as it contacts the nipple, seizes it with its lips and begins to suckle. Furthermore, the human newborn's mouth opens automatically when its cheek is touched just as the altricial chick's mouth opens or "gapes" automatically when the edge of its jaw is touched.

- ◆ The instinctual suckling program of the human neonate works in concert with its instinctual breathing program to prevent the milk from entering the windpipe. If a bit of food lodges over the windpipe, an instinctual coughing program is implemented that moves the offending food upward and out. The same kind of coordinated instinctual programs are also known to operate in many other activities of newborn babies—crying, vomiting, hiccupping, sneezing, urinating, defecating.

- ◆ The newborn is instinctually pre-tuned to judge accurately the healthy quality of the milk. If the milk is sweet, it continues suckling; if it does not taste right, it is instinctually guided to recognize the bad taste and to stop suckling immediately. Also, instinctual plans (not gravity) guide the movement of the milk from the mouth to the stomach. The act of swallowing by itself involves the coordinated contractions (timed in milliseconds) of more than a dozen muscles in the tongue, pharynx, and larynx and another dozen muscles in the esophagus.

♦ If the neonate is held so that its hands and knees touch the ground, it is instinctually programmed to move its diagonally opposite limbs and crawl. If it is held up with its feet touching the ground, it instinctually moves its legs and walks. If placed in warm water with its chin supported, an instinctual swimming program is implemented. If an object is placed in its palm, the neonate is instinctually programmed to grasp it. It is also programmed to follow moving objects with its eyes and to turn and look at the source of a sound.

♦ One of the most important discoveries in developmental psychology is that the newborn human (tested as early as forty-two minutes old) sticks out its tongue in imitation of an experimenter who protrudes his and also imitates many other human facial movements, such as mouth opening, lip protrusion, eyebrow movements, and smiling.[6] This recently discovered ability of the just-born human to imitate another person's patterned facial movements was bewildering to scientists. They realized it meant that, at birth, the human is programmed to perceive patterned movements on human faces and to translate them into similar patterned movements on its own face by manipulating in a highly coordinated manner the muscles of the face, tongue, and lips.

♦ Many additional instinctual programs guide the behavior of the human newborn, including unimaginably complex programs for functioning of each of the senses (vision, audition, olfaction, touch, pain, kinesthesis, taste, temperature), an instinctual program for rapid eye movement sleep, a program for the emotion of distress. Other instinctual programs, such as for exploration, play, and laughter, are implemented more gradually in association with maturation and social practice. With more extended periods of mat-

uration and practice, the instinctual guidelines or
programs are put into effect for each of the human
cognitive and mental capabilities including conceptu-
alizing, remembering, planning, and imagining.
Many other instinctual programs are set to begin on a
delayed or long-term maturational schedule; these
include a scheduled male voice change and initiation
of beard growth and a scheduled female program for
menstrual cycling and menopause.

In brief, the data converge on the conclusion that humans are
instinctually guided to do or learn to do everything that is
common to all *Homo sapiens* in essentially the same way that birds
are instinctually guided to do or learn to do everything that is
common to all members of their species.

Instincts of the Cuckoo Chick

◆

Writers who promulgate the theory that birds are simpleminded
automata usually provide one or more dramatic examples of
birds carrying out instinctual programs in a seemingly robotic
way. Thirty or forty years ago this theory was typically exempli-
fied by describing very young ducks, geese, chickens, or other
precocial birds rigidly, irreversibly, and automatically following
the first moving animate or inanimate object they perceived.
However, since this imprinting phenomenon has been shown
over the years actually to be flexible, nonrigid, and reversible, it
is no longer emphasized as an example of avian automaticity.
The most often cited example during recent years is the behavior
of the recently hatched blind cuckoo chick, which is said to act
robotically when it instinctively pushes all of the other eggs out
of the nest. Let us look closely at this dramatic example.

The cuckoo mother abandons her eggs and flies to warmer
climes after she deposits each egg surreptitiously in another
bird's nest. The cuckoo chick typically hatches first, before the

other eggs in the host nest. The only way the fast-growing cuckoo chick can obtain sufficient food from its foster parents is if it is the only chick in the nest. Within eight to thirty-six hours, it becomes the sole nestling by carrying out an instinctual program that has a number of components.[7] Beginning within hours after hatching, the blind cuckoo chick uses its sensitive wing tips to learn the dimensions and contours of the nest and the physical characteristics and locations of the other eggs. Subsequently, it wiggles about at the bottom of the nest until it maneuvers a host egg onto its unique back (which has a depressed concave area into which the egg fits). The cuckoo chick holds the egg in its back depression with its upraised stubs of wings, moves backward to the side of the nest, balances the egg on its back while pushing it to the top of the nest, and heaves the egg out of the nest while typically exercising sufficient skill to avoid tumbling out with the egg. After the egg has gone over the edge, the cuckoo chick rests. Then it repeats the same procedures, with intervening rest periods, until it ejects all the eggs from the nest.

This remarkable behavior of the cuckoo chick is as instinctually guided as the equally remarkable instinctual behaviors of the human newborn that were summarized earlier. To insure its survival, the cuckoo chick carries out complex instinctual activities that result in the removal of competing eggs from the nest. In the same way, to obtain needed nourishment, the human newborn at its first feeding carries out complex instinctual activities that result in its feeding at the breast—pushing toward the breast with its legs, searching with its arms, flailing from side to side, clutching the breast with its hands, seizing the nipple with its lips, and sucking vigorously while coordinating its breathing and swallowing.

On closer inspection, the behaviors of both the newborn cuckoo and the newborn human seem to blend sensible instinctual programs for survival with equally sensible moment-to-moment actions that implement or actualize the programs. The cuckoo chick quickly learns the dimensions and contours of the nest and the eggs. It removes each egg with increasing efficiency as it learns from feedback information to make just those precise coordinations needed to maneuver the egg on its back to the edge and out of the nest without itself falling over. (In fact, cuckoo chicks occasionally tumble out with the first or second egg they are ejecting, before their skills are perfected with practice.) In essentially the same way, the human newborn implements its inherited survival program for obtaining nourishment by moment-to-moment actions from which it learns quickly to suckle more and more effectively and efficiently. Instinct is thus blended with intelligence, or, more precisely, innate programs are implemented by moment-to-moment actions that require choices or decisions, which in turn require intelligence. Let us now turn to the very important topic of the intelligent implementation of instinctual programs.

Intelligent Implementation of Instincts
◆

To insure survival, certain instinctual programs have to be actualized immediately at birth. The instinctual human program that guides the newborn to feed from the breast must function efficiently from the start even though implementation of the program will be improved and fine-tuned with practice and learning. Most instinctual programs, however, are *not* constrained by survival needs to function in a fully satisfactory manner at birth. Consequently, most instinctual programs can be shaped and sharpened significantly by feedback, practice, and learning to fit closely the demands of particular circumstances. The flexible implementation that is needed to fit constantly changing conditions requires choice (judging or deciding between alternatives), which in turn requires intelligence.[8]

Although humans inherit programs for universally human behaviors such as talking, dreaming, and reasoning, how each individual talks, dreams, and reasons, and what he or she talks, dreams, and reasons about, are dependent upon experiential learning, accumulated knowledge, moment-to-moment feedback, and countless choices among alternatives made by application of intelligence. In essentially the same way, as we shall see in subsequent chapters, the instinctual programs of birds — programs for interspecies' communication, song production, nest construction, and navigation—are implemented flexibly and intelligently.[9]

CHAPTER FOUR

Avian Languages

Recent research has demonstrated that birds, much like humans, use both a body language and a vocal language to communicate everything that is significant for their way of life in their particular niche. Let us glance first at our own subtle body language before we turn to the communications of birds.

Human Body Language

◆

In addition to communicating via their obvious vocalizations, humans also communicate much meaningful information to each other by their subtle, unconscious body language of nods, gestures, facial expressions, postures, eye movements, and eye expressions. Examples of human body language include:

"I am submissive" communicated by a stooping posture.

"I am resolute and confident" communicated by a firm walking step.

"I am perplexed" communicated by head scratching.

"I am not concerned" communicated by shoulder shrugging.

"I am dejected" communicated by depressed lower lip with eyebrow frown.

Very few humans are aware of the tremendous communicative role of their own subtle body language.[1] Also, few realize that by using American Sign Language (or a comparable hand-gesture language), deaf people communicate as proficiently with each other as do hearing people speaking English or another language. Also, in the not too distant past, a very simplified hand-sign language provided a satisfactory means of communication among the American Plains Indian tribes who spoke different languages. The extent to which meaningful human communication can take place solely by body language is also seen in pantomime, a theatrical form dating back to antiquity in which actors communicate to their audience meaningfully without words, telling a story and expressing actions and emotions by skilled use of body movements, facial expressions, and gestures.

Avian Body Language
◆

Meaningful communication also occurs among birds via "eye-talk," mouth motions, raising particular feathers, stretching the neck, crouching, leaping, fluttering, and other movements. Although each bird species has its own body language, many species interpret certain movements in essentially the same way. For instance, various species interpret pointing the beak upward as "I intend to fly"; moving the crest downward often means "Watch out!" or "Danger!"; a number of species interpret raising the tail feathers as "I am threatening you"; and flashing a brightly colored crown patch is often translated as "I am ready to attack." Avian "eye-talk," the changing "look" in birds' eyes, can communicate both negative feelings or attitudes, such as dislike and resentment, and also positive ones, such as pleasure, enthusiasm, or curiosity.[2]

Birds also have a "beak language." During an eleven-year project with many birds that freely entered and exited from her open cottage, Len Howard discovered that birds also communicate by subtle beak movements that are, at times, accompanied by subtle sounds. Among several avian species, quickly opening and shutting the beak means "I want food," opening the beak slightly and hissing means "I am angry," and opening the beak halfway with a faint sound (like a bottle uncorking) means "I am disgusted."[3] Howard concluded from her close observations of many kinds of birds that "My birds can make me understand by their notes and actions or inflections of movement what is passing through their minds. If I do not get their meaning at once they soon find a way of communication, often by pantomime."[4] Other observers who formed personal relationships with birds living freely in their open households similarly concluded that birds can effectively make their concerns known to humans by facial and eye expressions, particular body movements, sounds, and pantomime. One such case[5] shall be described in detail in Chapter 5 and other cases in Chapter 8. The amount of data on avian "visual display" (body language) has reached the point where even exceptionally "hard-headed" scientists, who adhere to the official dogma that birds are machines, concede that the avian visual mode of communication is a vast, unmined area and that deciphering or translating it remains one of the greatest challenges in the study of avian behavior.[6]

Avian Call Language

◆

The body language of birds is one component of a larger communication repertoire that includes three distinguishable types of meaningful vocalizations: short calls; longer, more musical songs; and social chattering and muttering, which appear at times to resemble conversation.

The short sounds or calls take various forms, including whistles, hoots, squawks, rasps, and drummings. Various carefully observed species have distinct sounds or calls for their common feelings or emotions (anger, contentment, distress, surprise), for many activities ("I have found food," "Attack!" "Victory!"), for several kinds of courtship activities, and for more than one kind of danger. The calls have many social uses: They play an important courtship role in mutual stimulation of mates; they synchronize many pair activities; and they are important in coordinating the behavior of the family group or flock in feeding, avoiding predators, flocking, and migrating. Calls also serve as "signatures"; each bird identifies or "names" itself with its calls. Although some calls clearly express feelings and emotions, most appear to transmit particular information, such as "I have just found a large amount of food."[7]

A substantial proportion of bird calls appear to be innate and are understood by conspecifics (birds of the same species) and, at times, by birds of other species. However, many bird calls are learned. Researchers have succeeded in distinguishing about three hundred different crow calls. Scientists have satisfactorily deciphered only some; for instance, five crow calls are known to refer to different kinds of danger. Many "linguistic" complexities have already been found in these crow calls: There are local dialects among crows in the United States; isolated groups of crows do not understand some calls of "mainstream" crows; and crows on both sides of the Atlantic understand some but not all of each other's calls.[8]

Human understanding of avian call language is just beginning. As research proceeds, we can expect more surprises similar to the following recent unexpected discoveries.

It has been found that humans are deaf to some dimensions of bird calls. Humans hear neither the high-frequency sounds that a few avian species can hear nor the low-frequency sounds that many avian species can hear. Also, birds far exceed humans in the ability to separate sounds arriving in very rapid succession. Surprisingly, the time resolution of birds has been estimated to be nearly ten times better than that of humans; it appears that people hear as one note what birds hear as ten separate sounds![9] (Birds' speeded-up temporal resolution is consistent with how they experience other dimensions of time. As Howard writes, "Time goes at a different pace to birds, they have faster pulses, higher temperatures, their sight and hearing quicker and their actions often at a pace we cannot follow with our eyes."[10]

The "same" bird call can mean different things depending on the context. An example is a particular call of male laughing gulls, which has a different meaning when it is emitted in the presence of an intruder, a prospective mate, or an actual mate.[11] Similarly, in different contexts, a goose's honk informs about the presence of danger, the choice of a resting place, the presence of a good grazing site, the characteristics of the weather, and when it is time to take flight and depart.[12] The situational interpretation of calls also plays a role in human communication, for instance, when humans utilize one sound, such as the sound of their automobile horn, to say in various contexts "Watch out, I'm passing you," "You forgot to put your lights on," "Thank you for letting me pass you," "You just went through a red light," "Just married," and "Hello."[13]

It has been reliably observed that birds at times purposively use calls to deceive other birds. For example, when bird A is chasing a desirable insect, bird B gives a false alarm that causes bird A to hesitate briefly, whereupon bird B grabs the prey.[14]

Avian Song Language
◆

In addition to their body language and call language, about half the bird species also utilize a third mode of communication — melodious songs. The song of a male bird informs another male of the same species that "This is my territory"; it informs the appropriate female that he is of the same species and is interested in mating; and the quality of his song informs both males and females that he possesses particular levels of confidence, competence, vigor, and cheerfulness. Also, the songs of birds betray or reveal their emotions—including joy, contentment, love, and sadness. We shall return to this topic when we focus on the avian musical sense in Chapter 6.

Avian Conversations

◆

Ornithologists seem to assume that birds communicate only by their calls, songs, and "visual displays" or body language. Two additional possible means of avian communication have been totally neglected: the chattering that is often heard among large flocks of birds and the quiet murmurings and whisperings that have been observed among pairs of birds. No one has investigated the chattering of flocks to determine to what extent and in what way the participants are communicating meaningfully. With respect to the murmurings and whisperings of avian pairs, however, recent data suggest a conclusion that, when rigorously confirmed, will shock human beings: Two birds may at times literally converse with each other in essentially the same way people carry out a conversation, that is, by taking turns speaking, waiting for the other to finish, and relating their statements to the "topic" of the conversation or to what has been said previously. Data supporting the seemingly incredible notion of avian conversations include the following:

A surprising discovery was made when crows' vocalizations were recorded via microphones secretly placed where they ate and where they rested. The crows vocalized many whispered sounds that investigators had never heard before. Analyses of these sounds suggested to the astonished researchers that the crows may have been actually conversing and, in fact, may be "great conversationalists."[15]

An investigator who succeeded in monitoring the sounds of a pair of nesting blue jays over an entire season was surprised at the many unexpected vocalizations that were emitted at regular times. According to the researcher, the sounds indicated that they "hold conversations . . . as if they were reviewing the day's events."[16]

A professor of zoology rescued a nestling great horned owl buried in the snow, raised it to maturity, and maintained a close friendship with it after it was set free. This careful scientist dared to report that "It is the many varied soft and hushed sounds that [the owl] makes that I find most fascinating. I hear them only when I am next to him; they are his private sounds, reserved for intimacies."[17]

Other investigators have published similar reports of mated birds, and a few unrelated birds, taking turns vocalizing; the bird spoken to gave its full attention to the speaker and never vocalized at the same time, as if the two were holding a conversation.[18]

Researchers and scholars who have studied the data on avian communication carefully write that (a) the communication code of birds such as crows "has not been broken by any means,"[19] (b) "Probably all birds have wider vocabularies than anyone realizes,"[20] and (c) greater complexity and depth are recognized in avian communication as research progresses.[21]

Humans may have understood the communications and possible conversations of birds no better (or worse) than birds have understood the communications and conversations of humans. I expect that further research with birds living free in nature and also with bird pairs living freely in human households "as part of the family" will demonstrate that particular birds literally converse — take turns exchanging meaningful information. Also, when researchers realize that even now, and certainly before the television age, there existed many illiterate families all over the world with very small vocabularies, they can proceed to test the revolutionary hypothesis that some birds communicate to other birds as much information as some ordinary people communicate to other people.

◆

CHAPTER FIVE

Lorenzo, the Communicative Jay

All official ornithology texts share a very serious deficiency.[1] They discuss the behavior of birds in general and of particular avian species, but they do not recognize that birds are individuals and they never focus on the behavior of one particular individual bird. This book remedies this deficiency by looking at birds not only at the class and species levels but also at the level of particular individuals. To put flesh and blood on the implicit themes of Chapter 4—that birds can communicate everything significant in their lives and that there is much more that they consider significant than official science ever imagined—we will now focus on the communicative interactions of one particular closely observed bird.

A very young, injured scrub jay that had apparently fallen out of its nest was brought to Robert Leslie, a naturalist who, together with his wife, operated a clinic for orphaned wild animals.[2] They nursed the tiny bird with an eye dropper, feeding its voracious appetite, and restored it to health. This male jay, Lorenzo, lived freely in the Leslie household "as part of the family" for nearly three years (when he left to start a new life nearby with a mate). While he lived with the Leslies, Lorenzo had an open cage that he could enter whenever he wished. Later, a large aviary was built for him outdoors, which was also kept open so he could fly in and out. (Occasionally he was locked in his cage or aviary either for his protection or for punishment.) After a period of adjustment, he spent his days partly indoors and partly outdoors; each evening for about half an hour he literally socialized with the Leslies in their den; and at night he slept either in his indoor cage or on a particular lamp shade.

Lorenzo communicated to the Leslies everything that appeared to matter to him by special sounds and by body language, which included movements of each part of his body supplemented by changing eye expressions. Leslie reported that "the intensity and exactness of Lorenzo's ability to communicate impressed human observers. The changing look in his eyes clearly revealed curiosity, in-depth study, likes and dislikes, enthusiasm, feelings of abuse, resentment, restless boredom, and once in a while, downright hatred. These were the everyday expressions he made known."[3] His body language included such statements as the following:

"Let me out" [of the cage or aviary]—a particular kind of pecking and gargling

"I want a drink"—a metallic throat sound together with a particular kind of head bobbing

"No more food"—a backing-away movement with a raising of the head and a peculiar turning motion

"I'm mad at you"—a particular kind of hand pecking

"Give me that"—another kind of pecking

"I'm sorry"—a light nibbling, typically on the person's hand

"I don't want to come"—a peculiar call given only in those rare instances when the Leslies called him and he did not immediately come to them because he was doing something he apparently viewed as more important. If the Leslies persisted in their calling after Lorenzo gave his "don't want to" call, Lorenzo always gave another unusual call, which Leslie interpreted as "the jaybird equivalent of 'Go to hell!' "

Similarly, by various combinations of sounds and actions, Lorenzo communicated such desires as "Place my bathtub here" and "I'm ready to play now, how about you?" He used a unique call to say "Good-bye" when he departed under ordinary circumstances but not when he "sneaked" out to perform a forbidden act or when he left because his pride had been hurt. Some communications, which he used only once, were obviously created right on the spot. For instance, he told Leslie "Come and help" by particular "hollering" sounds, plus pulling on Leslie's chest hair and skillfully leading him to a baby bird that (like himself at an earlier time) had fallen from its nest.

Lorenzo's understanding was surprising to the Leslies and to other people who interacted with him. He definitely understood that human laughter can have more than one meaning, since he became angry when people laughed at him when he was behaving seriously but was pleased if they laughed when he acted the role of a clown. He also understood the concept or idea of "ownership," since he kept careful track of two dozen toys he owned. (An example of his toys was a toilet paper core that he used as a scooter by standing on it as he propelled it along with the skillful use of his beak, wings, and talons.) He checked daily to see that each toy was where he always placed it, and if one was missing he complained immediately by wailing and pecking on Mr. or Mrs. Leslie's hand until he regained it.

Lorenzo understood the concept of "a game with rules" since he skillfully guided the Leslies to play three kinds of hide-and-seek games with him, each with different rules—they hid one of his toys and he searched for it, he hid while they searched for him, or they hid while he searched for them. Astonishingly, he organized particular kinds of "chase" games that included a number of neighboring birds of various species. For instance, in one game the rule was "Chase that bird which is carrying the aluminum ring [like a relay baton] in its beak." Lorenzo also invented other games and, with practice, became increasingly skilled at each.

Lorenzo understood the idea of "singalong" since he listened silently to classical music but imitated and sang along with soloists (such as Elvis Presley and Joan Baez). He demonstrated that he understood the meaning of "socializing" by joining the Leslies for about half an hour each evening and engaging in continuous social activities, such as cuddling against Mrs. Leslie's neck and kissing her, combing Leslie's beard by using his beak and talons, perching and cooing on their shoulders, "soft-talking" or whispering to them while his eyes were half closed, and, at times, attempting to tell them via his vocalizations and body language about an exciting event in his day. (The Leslies clearly understood the latter pantomimelike communications if they knew the context—that is, if, earlier in the day, they had witnessed the exciting event, such as when two dogs had knocked down his aviary.)

When locked in his aviary for punishment, Lorenzo would offer the Leslies a valued toy for his release, thus indicating behaviorally that he understood the concept of "bail." He also understood the idea of "trading" since he first tugged at a visitor's ring, bracelet, or jewelry, then quickly went to get one of his toys and, after returning, gave the toy to the visitor while tugging again at the item desired. Visitors "traded" their valuables for Lorenzo's toy, believing he just wanted to borrow it and would surely return their property to retrieve his toy. However, Lorenzo apparently wanted the visitors' valuables more than his toys because he never returned their valued objects. The Leslies always had to retrieve and return them.

Lorenzo also understood the sleight-of-hand art of the

pickpocket. If he did not acquire a visitor's valued item by "trade," he got it by following the basic working principle of the professional pickpocket—quickly take the valuable when the victim is distracted. He tweaked the earlobe of one victim, mussed up the hair of another, and placed an after-dinner mint down the collar of a third. When each victim was distracted, he swooped down and yanked away the desired object.

Lorenzo's behavior indicated that he understood the concepts of "help" and "share." He shared his food, sleeping cage, and toys for two months with a maturing male bird of another species that he had led Leslie to rescue. He was also repeatedly observed in the surprising activity of giving his own food to a mother squirrel with a hungry brood of young. He indicated behaviorally that he understood the meaning of "secret revenge." When he saw two crows (who had been pecking him and taking his food) leave their nest, he quickly flew to their nest and rolled out the eggs. He then flew back into his cage, called for Leslie to lock it, and pretended to be asleep when the crows arrived later looking for the culprit.

Lorenzo's communicative behavior included both clearly interpretable aspects and also subtle aspects that remain to be definitely interpreted. There was no doubt that Lorenzo communicated meaningfully with the Leslies via vocalizations and many dimensions of body language, including "eye-talk" and, occasionally, pantomime. Other aspects of his behavior that appeared to be communicative but are not so definitely interpretable included, for example, the following.

Lorenzo vocalized consistently in the same way in certain particular circumstances. For instance, his vocalizations were of a particular kind whenever he followed one or both of the Leslies around the house or garden while flying alongside or riding on a shoulder.

During the beginning weeks of Lorenzo's relationship with his mate, when she visited him regularly in his outdoor aviary, they appeared to "talk" back and forth as they continuously vocalized, chattering and "jabbering" with each other. Did

Lorenzo communicate meaningfully with his mate? There is much more to be learned about this kind of avian vocalization. In Chapter 8, I shall return to this topic and will describe in detail the communicative activities of other carefully observed individual birds.

◆

CHAPTER SIX

Avian Music, Crafts, and Play

A series of research projects show that birds, in essentially the same way as humans, possess a musical sense, a sense of craftsmanship, and an ability to play and to enjoy life. This chapter synthesizes these surprising findings.

Avian Music

◆

About half of the nine thousand or so species of birds are songbirds, some of which sing exceptionally beautifully. The eminent field ornithologist Alexander Skutch writes: "One must be aesthetically insensitive not to recognize the beauty of bird songs, nor to be emotionally stirred by them, especially if heard in their natural settings where they are most effective."[1] Ironically, just when leading scientists are beginning to understand that birds, like humans, possess a special musical-aesthetic intelligence, the habitats of songbirds are being destroyed worldwide at such an accelerated rate that soon few people will know what it means to hear a talented wild bird sing.[2]

The most intensive investigation of bird song was carried out by the English musicologist Len Howard, who devoted all her time during the 1940s to studying wild birds and their music while living with them in her open country cottage and observing them in the surrounding orchard, garden, and lawn.[3] Since she provided supplementary food (on a large table) for many birds and approached them with respect and a loving attitude, she became personally acquainted with many and knew some for their entire lives. She became even more closely acquainted with dozens of birds who entered and exited freely from her cottage after she opened all windows and doors during the warmer months and allowed sufficient openings during the colder months. Her intimate study of bird songs led to four surprising conclusions:

◆ Birds, like humans, enjoy their songs. They take pleasure in singing, and they enjoy hearing even their territorial rivals sing.
◆ Birds not only convey messages and express feelings and emotions in their songs, but at times they sing simply because they are happy.
◆ Conspecific birds can be reliably identified by their unique variations of the species' song. In fact, conspecific birds apparently differ in musical talent as much as humans. This unexpected variability is due to the individual bird's interpretation of the theme, his technical ability in executing it, his "style" of delivery, and the quality or timbre of his voice. Some very poor singers are found in every songbird species. For example, a few male chaffinches are unable to sing the finale, or terminal flourish, which is the highlight of their species' song. There are also very superior musicians among songbirds. For instance, over a period of a few days, a talented blackbird creatively and spontaneously composed the opening phrase of the Rondo in Beethoven's violin concerto. (He had *not* previously heard it.) During the remainder of the season he varied the interpretation of the phrase; "the

pace was quickened towards the end ... a rubato
effect that added brilliance to the performance."[4]
♦ Birds produce beautiful songs, not by robotically
carrying out a preset program but by individual talent, creativity, practice, and experience.

Howard's general conclusion, that bird songs have the basic
musical characteristics of human songs, is supported by the
noted philosopher and ornithologist Charles Hartshorne, who,
over a period of fifty years, recorded and analyzed the songs of
hundreds of avian species living on every continent.[5] Hartshorne
agrees that songbirds possess a musical sense, or "feeling for
music," that some of their songs are beautiful and musical in the
same sense as some human songs, and that sometimes birds sing
because they are happy. After spending nearly fifty years in the
tropics observing wild birds, Alexander Skutch adds that birds
surely must be conscious because only conscious beings can
appreciate the aesthetics of songs the way birds do.[6] Ethologist
Colin Beer agrees, writing that the songs of birds are often rich in
harmonic content and have theme, melodic transposition, melodic inversion, and other dimensions of formal musical organization that derive from an aesthetic sense of beauty.[7]

The sensory limitations of humans prevent them from
perceiving the richness of bird songs. Birds are roughly ten times
more proficient than humans in creating and also acoustically
separating out a very rapid succession of sounds. Therefore, it is
not surprising that recordings of bird songs played at reduced
speed (from one-half to one sixty-fourth original speed) reveal
new dimensions in their songs that humans had not previously
imagined.[8] A bird may perceive a song that lasts only two
seconds as a lengthy and intricate composition. For instance,
when played at one-quarter speed, a hermit thrush's two-second
song sounds like a human composition that has between forty-
five to one hundred notes and twenty-five to fifty pitch changes
and approximates a pentatonic scale with all the harmonic intervals.[9] These facts lead to three important considerations:

First, the songs of birds and humans are much more alike in

musical characteristics than previously imagined. The major difference between them is the much shorter time span of bird songs. However, within a brief temporal span, birds can discriminate and "pack in" a huge number of musical sounds.

Second, it is open to question whether the best human composers can create compositions packed within a time span of a few seconds that can musically surpass the songs of birds.[10]

Third, since songbirds have an accelerated metabolism, heartbeat, and respiration, live at a much faster tempo than humans, discriminate more stimuli within a given unit of time, and react faster to stimuli, they may experience in their few allotted years as much as humans do in their much longer life span.

Careful research has demonstrated that, in at least some species, male birds with more pleasing and more complex songs are preferentially selected by females and mate before their rivals.[11] Some unmated male birds, such as song sparrows, may spend as much as nine hours singing during the day. Even though they follow a "sparrow song pattern" (which identifies members of the species to each other), each bird sings its own individual songs. An individual song sparrow may compose about twenty different songs, a solitary vireo may compose about seventy-five, a winter wren about thirty, and a marsh wren as many as one hundred.[12]

If we focus on songbirds of one species, say the European blackbirds intensively studied by musicologist Howard over eleven years, we learn the following.

Each male European blackbird composes many original melodies that are typically beautiful, calm, and reposeful. Some of their songs are "in approximately correct classical notation of interval and time signature (using compound and simple time)" and are surprisingly similar to human classical music.[13]

Blackbirds use many of the same musical techniques as human composers. Howard writes:

Each bird appears to try for variety of effect by singing his tunes at different paces, adding embellishments, altering the key, singing in major and minor modes, using staccato [notes sharply detached] and legato [notes smoothly slurred]. The blackbird even tries turning his phrase upside down. All these ways of treating a tune are in the human composer's technique. He varies phones and has a wide range of intervals in his songs. His delivery is calm and controlled. Vibrato [or tremolo] is used effectively and with great discrimination.[14]

Young male blackbirds improve their skills by listening attentively to talented older birds. Since they differ widely in singing ability, in ability to compose tunes, and in readiness to take or "borrow" the tunes of other birds, each blackbird's song repertoire is very individualistic and no two blackbirds sound alike. The typical male blackbird sings many different tunes in different rhythms strung together with only a slight pause between each tune.

Songbirds, like human musicians, aesthetically utilize changes in tempo—*accelerando* (gradually increasing tempo), *ritardando* (gradually slower), and *tempo rubato* (lengthening certain notes while correspondingly shortening others). Songbirds alter the sound of their notes without altering the timbre in much the same way human soloists alter their words while keeping the same quality of timbre or voice. Both birds and humans use rhythm as the foundation of their songs.[15] Hartshorne emphasizes that songbirds, like humans, manifest musical sense when they improve their songs with practice and also when they "steal" or imitate the songs of other birds.[16] In fact, the musical specialty of the mockingbird (imitating the calls and songs of virtually all of the birds in its vicinity) requires a superb ability to distinguish and reproduce the subtle variations in loudness, duration, and timbre of each note.

In addition to proclaiming "Here I am with my territory," the songs of male birds at times seem to resemble "love songs." For instance, a male linnet, carefully observed by Howard, sang in "a mysterious musical language of phrases half-whispered, half-sung, intermingled with bell-like trills and wistful sounding, caressing notes" while he continually faced his mate as she built her nest, brooded her eggs, and "often paused in her work to look up at him." (This well-mated pair had a long acquaintance with Howard, did not fear her, and thus behaved naturally in her presence.)[17]

Essentially like humans, songbirds use their musical abilities in various ways, not just for proclaiming self and territory or for courting or expressing affection. They sing at times simply because they are happy, or to attract the attention of another bird or person, or to entertain their flockmates. Howard writes that the entertainment function was surprisingly common among the great tits she observed carefully. She writes:

> Watching the bird singing his imitation [of the song of another species], he occasionally seems [to be] doing it for fun and his performance may have an audience of several other birds who hover around, gazing at him while he repeats the imitation many times running, with head and neck outstretched in humorous attitude, and in dancing manner hopping backwards and forwards from one twig to another. He appears to be playing the fool and enjoying himself immensely while the audience watches with interest. This happens in autumn when Tits mix together amicably, and often invent ways of passing the time. Once this performance is over the audience disperses and the imitation is not again repeated.[18]

The enchanting phenomenon of group singing was also described in detail by Howard. In a typical instance, about thirty linnets arrive at a particular tree

> in an excited, purposeful manner, all alighting simultaneously and near together. Then one bird starts singing, by about his third note another joins in, the rest following in quick succession, perhaps six at a time entering the chorus, some trilling, some twittering, some singing

upward and downward slurred notes until all voices combine in a great crescendo held for a few moments. Then the volume of tone gradually sinks while over the fields another flock is seen heading in rushing flight for the Linnets' tree. They settle beside the singers and now the chorus again swells, the two flocks by degrees uniting in song until every bird has joined in a redoubled crescendo. The sound of this chorus travels far, stirring other flocks feeding in more distant fields. Before long there is another hurried rushing of wings, and the tree, bare of leaf, becomes laden with Linnets on every twig, all voicing their loudest fortissimo. The song reaches its climax which is held for a time, then the tone falls and rises again, until suddenly, all together, the birds take to flight, their minds tuned to one accord in song now act under one impulse to fly.[19]

Similar performances are seen in other species. For instance, a large flock of goldfinches has been seen to land together on a tree and "make music together in sympathetic accord," each bird singing its own song in unison with all the others and all pausing and stopping together simultaneously.[20]

Five surprising discoveries in sequence showed that birds, like humans, can sing duets, trios, quartets, and quintets.

First it was discovered that the songs of bou-bou shrikes are actually duets produced by the precisely coordinated singing of a mated pair.

Next it was discovered that each pair composes a large number of duets. Some are composed antiphonally; that is, each bird takes turns coming in with the next note (or sequence of notes) with incredible precision (within a tiny fraction of a second). The remaining duets are composed polyphonically — the two birds simultaneously sing the same or harmonized notes.

Next came the shocking discovery that some songs thought to be duets were actually produced by a trio or even by a quartet or quintet. The third, fourth, or fifth birds, who joined in the singing with split-second timing either antiphonally or polyphonically, included the grown offspring of the mated pair and, at times, conspecifics from neighboring territories.[21]

Further research revealed that at least 222 species of birds (members of at least 44 bird families) sing duets, trios, and so on.

Finally it was reliably observed that duets occasionally were sung by two birds belonging to totally unrelated species.[22]

Distinguished scientists such as Howard Gardner have presented very powerful evidence that there are many kinds of intelligence, including linguistic intelligence, spatial intelligence, social intelligence, and musical intelligence.[23] The scientific data lead to the conclusion that songbirds, like humans, possess musical intelligence.

Avian Aesthetic Sense
♦

A series of reliable observations converge on the conclusion that birds and humans have a similar sense of beauty.

As judged by Howard, Hartshorne, and other musical experts, birds make their songs more beautiful as they revise them.

The displays of the peacock and of the male Argus pheasant make sense only if the courted females appreciate their beauty.[24]

Birds have demonstrated their aesthetic sense in the laboratory by choosing symmetrical designs and rhythmic patterns in preference to irregular ones.[25]

Various kinds of bowerbirds exhibit a marked appreciation for what humans consider to be colorful, pleasing, and beautiful. They decorate their bowers with gorgeous flowers, beautiful feathers, brightly colored berries, and iridescent objects. When these objects lose their luster or beauty, the birds replace them with other beautiful objects.[26] Reliable observers have reported that, when constructing their displays, bowerbirds at times act like human painters who critically examine their canvases or like people who are trying out many rugs in their homes to decide which ones to buy and where to place them.[27] The behavior of one carefully observed male bowerbird (an orange-crested gardener) is described thus:

> Every time the bird returns from one of his collecting forays, he studies the over-all color effect. He seems to wonder how he could improve on it and at once sets out to do so. He picks up a flower in his beak, places it into the mosaic, and retreats to an optimum viewing distance. He behaves exactly like a painter critically reviewing his own canvas. He paints with flowers; that is the only way I can put it. A yellow orchid does not seem to him to be in the right place. He moves it slightly to the left and puts it between some blue flowers. With his head on one side he then contemplates the general effect once more, and seems satisfied.[28]

Avian Craftsmanship

◆

Using only their feet and their beaks, birds are able to construct secure and durable homes that are protected from predators and from climatic extremes. A high level of craftsmanship is typically involved since many of these nests can be duplicated only by expert human craftsmen who are exceptionally skilled in using special tools.[29] The nests are strongly built from materials that are skillfully integrated and firmly secured to hold the entire family of parents and restless chicks as well as to survive heavy rains, strong winds, and wide swaying of the bough. Nests are proficiently and wisely constructed: The thickness of the walls correlates with the weight of the chicks to be reared; many sensible adjustments are made related to the position of the supports; protection from strong winds is provided by a roof covering or by especially strong supports. Protection from rain is also assured by one or more contrivances—by building a special ceiling or roof or by using either waterproof material or a porous material that drains readily. To understand the skill involved in avian craftsmanship, let us look at six representative types of avian nests.

The simplest nests, such as those of crows, are strong, snug, safe, and highly functional. They are constructed by skillfully working each twig into an appropriate place and lining the nest with mud and soft material. The nest may take from five to thirteen days to complete. At times it is constructed by a cooperative effort, with two or three yearlings bringing in sticks, mud, and fine lining materials, while the breeding female arranges the sticks and lining, and the breeding male typically stands guard.[30]

Male weaverbirds construct intricate nests of great density and durability by using the idea of warp and woof while skillfully carrying out the functions of both basket weavers and of weavers using looms. Not surprisingly, a lengthy period of practice is required for the weaverbirds to develop their inherited motor propensities and to skillfully integrate them with the moment-to-moment judgments they use in weaving.[31]

The tailorbird first searches for material that can serve as thread (such as cotton fibers, pieces of string, spider's silk, or bark fibers). It then builds its home by sewing leaves together — piercing holes at the edges of the leaves with its beak, drawing the threads through, and knotting them on the outside so that the stitches hold while concomitantly maintaining the leaves in juxtaposition by skillfully using its beak and feet.

Combining the techniques of both basket weavers and weavers of Oriental rugs, penduline titmice construct exquisite nests that are so durable and beautiful that children in Eastern Europe are delighted to wear them as slippers and the Masai tribes in Africa are happy to use them as purses.[32]

Ovenbirds skillfully mix about two thousand lumps of clay with a binder (of plant debris, cow dung, and straw) to construct their nests, which look like miniature adobe huts or old-fashioned domed bakery ovens. The nest, which has a foyer separated by a small connection from an inner nesting chamber, is built so that a human hand cannot reach the eggs in the inner chamber.[33]

The nests of African hammerheads are large ball-shaped structures, about two yards in diameter, that are constructed with sticks and lined with clay. They are strong enough to support the weight of a person standing on the domed roof, have a protective entrance so small that the birds must creep and squeeze in, and contain three separate rooms — an upper sleeping and incubation room, a middle room for the young when they have outgrown the upper room, and a vestibule or lookout room. Some species of bowerbirds build equally remarkable bowers reaching nine feet in height and resembling a thatched hut with several rooms and with a surrounding circular lawn.[34]

Birds inherit the readiness to build a certain kind of nest that is refined and perfected by learning through practice in essentially the same way as humans inherit the readiness to speak and to understand language. Just as human infants have instinctual "expectations" of the underlying structure of human language and the ideal pattern of human speech, birds have instinctual "expectations" of what the complete nest should look like.[35] Birds build safe and sturdy nests by using their inborn readiness to construct a nest, their inborn conception of the ideal nest, and moment-to-moment intelligent application of their beak, legs, and other body parts to choose, gather, and skillfully interblend appropriate natural materials.

Avian Fun, Play, and Dance
◆

As indicated by the behavior of the jay Lorenzo in Chapter 5, birds, like humans, can play and have fun. Their play varies from simple enjoyable activities to complex organized games. Many kinds of birds have been reliably observed to play with sticks, leaves, feathers, pinecones, and many other objects; to repeatedly drop a small object in midair and catch it before it hits the ground; to ride down a rapid tidal current or to slide down a snowbank feet first, walk back up, and ride or slide down again apparently for the sheer joy of it. "Chase" games, "follow-the-leader," and "cat-and-mouse" games with quarry are also quite common.[36] In fact, a group of closely observed birds of various species devoted a surprisingly large part of their day to playing "chase" with birds of their own and other species and indulging in sunbathing, flying, and singing as if they were recreational activities.[37]

Dancing has been documented in ostriches, manakins, cranes, and other birds that vary their steps in rhythmic patterns while they turn, spring, strut, and jump. Cranes, for instance, perform "solemn and stately dances"; sometimes males dance while females observe, sometimes a male and a female dance, and sometimes the entire flock dances rhythmically together. The dances are often held in areas that the birds have energetically cleared to use just for dancing.

Although birds at times dance simply for pleasure, dancing also plays an important role in courtship. Male manakins of various species "perform astonishing courtship dances" as they dance in pairs or in groups. The males act like whirling dervishes and perform acrobatic aerial displays and juggling acts (replacing one another like juggling balls) while one or more females observe. Surprisingly, the formal dances of various human cultures are closely patterned after the dances of native birds. The Bavarian peasant Schuhplatter dance imitates the dance of the native black grouse with the men dancing and the women rotating in a fixed position watching the men. The Blackfoot Indians imitate the dance of the sage grouse, the Jivaro Indians mimic the dance of the cock-of-the-rock, and the Chukchee have a dance patterned after the dance of the ruff.[38]

◆

CHAPTER SEVEN

Avian Navigation

If modern human navigators suddenly found they were without instruments over an unfamiliar area, they would not have the slightest idea how to return home. Nor would they be better off if they had a compass because they still would not know the direction of home. To return home, they would need a compass, a very precise clock, and a sextant together with astronomical tables and sophisticated navigational knowledge. Modern humans require all of these instruments to calculate their latitude and longitude and the direction to take to return to the latitude and longitude of home—provided they know these home coordinates.

Migratory birds are far superior to humans in their ability to navigate and return home without instruments. Great shearwaters find their nest on the tiny island of Tristan da Cunha in the middle of the South Atlantic even though each bird begins the journey from a different part of the north European coast and flies more than seven thousand miles over the Atlantic to reach home. Similarly, sparrows succeeded in returning to their nests and nestlings in San Jose, California, after they were experimentally displaced to either Louisiana or Maryland. Manx shearwaters found their way back to their nestlings on a tiny island off the coast of Wales after they were displaced to Boston or Venice, and Laysan albatross returned to their nests on Midway Island in the middle of the Pacific when they were moved to Japan, the Philippines, or the western coast of the United States.

Prior to the recent explosion of sophisticated research, scientists believed that birds required no special awareness or intelligence to perform their migrations and their navigational and homing feats. These performances were viewed as automatic activities, possibly involving an unknown special sense, such as sensitivity to the earth's magnetic field. No one ever considered the possibility that birds might navigate much more easily but in essentially the same way as Polynesian seafarers who sail without instruments over the Pacific by observing, analyzing, and synthesizing subtle cues that are available in nature.

The earlier view that avian migration, navigation, and homing are more or less automatic was shown to be clearly inadequate a few years ago when accumulated research showed that, in addition to performing the difficult task of correcting for displacement (by storms, winds, mountains, and other hindrances), birds integrate an astonishing variety of celestial, atmospheric, and geological information to travel between their winter and summer homes. As we shall see, birds sensibly utilize many of the same natural cues as Polynesian navigators, including landmarks, wind patterns, star patterns, and the apparent movements of the sun. In addition, birds are able to use a variety of informational cues that are not readily available to humans, including infrasounds, ultraviolet light, polarized light, barometric pressure, subtle olfactory cues, and variations in the earth's magnetic field.[1] Leading research specialists in avian navigation, such as R. Robin Baker, Stephen Emlen, and Charles Walcott, have interpreted the accumulated data as showing that birds have "a whole set of very sophisticated guidance systems,"[2] use "a whole array of navigational strategies,"[3] and "seem remarkably like people [since] they use a variety of cues in their travels ... [and] may even use several cues simultaneously, checking one against the other."⁴ The ability to use a variety of cues is very advantageous especially because any particular cue can disappear suddenly—landmarks may be hidden by fog, the sun and stars may be obscured by clouds, the direction of the wind may change suddenly, and disorienting magnetic anomalies may lie over the flight path.⁵

In brief, one of the most surprising and important discoveries of recent scientific research is that avian navigation, in essentially the same way as much human behavior, is characterized by the ability to gather a variety of informational cues and to interpret and coordinate them so as to move closer toward a goal. Let us look at the cues birds use to migrate, navigate, and home.

Using the Sun as a Compass

♦

An impressive discovery of modern ornithology is that birds, like Polynesian navigators, use the sun as a compass to determine direction by taking account of its apparent movements across the sky. A series of rigorous investigations established the following facts.

Both homing pigeons and birds that migrate during the day use the sun as a compass.⁶ (When the sun is obscured, they turn to other natural compasses, as discussed later.)

Using the sun as a compass is a difficult feat for both humans and birds; they must compensate for the sun's changing position in the sky throughout the day, for its very slow rate of movement at certain times (near dawn and dusk), for its very fast movement at noon, for its different position or arc through the sky on different days even at the same location, and for its different rate of movement at different latitudes.⁷

Birds are able to head consistently in one direction because they understand how the changing position of the sun throughout the day is related to direction.⁸ They have an innate predisposition to attend closely to the sun and to observe its movements; from these observations they *learn* early in life the relationship of (compass) directions to the sun's apparent movements and the passage of time.[9]

Since the sun is the most striking feature in the avian outdoor environment, birds presumably become aware of its visible movements and regularities as they continually experience its light, its heat, its direct blinding brightness, its shadows, its rising, its setting, and its constant movement across the sky. As they repeatedly experience the regular twenty-four-hour daily cycles of light and dark, birds also become aware of the regular passage of time and are thus able to set or calibrate an "internal clock."

"Reading" the Stars
◆

Just as day-migrating birds attend to and learn early in life the movements of the sun and later use this learning to determine directions, so do night-migrating birds attend to and learn in early life the patterns of the stars and later use this knowledge to head or orient in a particular direction. One set of experiments with several warbler (*Sylvia*) species showed that these nocturnal migrants were using star patterns and also, possibly, apparent star movements to find and maintain direction in essentially the same way as intelligent humans might use these cues.[10] A second set of experiments, with a variety of bird species including waterfowl, shorebirds, and a large number of songbirds, demonstrated that many kinds of birds can select their migratory direction when the stars are visible and that they are unable to do so or their orientational accuracy decreases under overcast night skies.[11] A third set of experiments with indigo buntings yielded three important findings.[12]

◆ These birds attend to the night sky when they are very young and, at that time, they learn star patterns and particular useful facts about stars—for instance, that the stars around the North Star rotate the least.

◆ By learning this information, the young birds acquire the equivalent of a compass—a fixed, unmoving north.

◆ The instinctual predisposition to attend to stars and learn their patterns is implemented individualistically. Contradicting the notion of birds as automata, each individual bird learns to determine direction by selecting and using a different star pattern in the vicinity of the North Star.

To compensate for displacement and construct the mental equivalent of a map, navigating birds utilize other features of the natural environment, to which we now turn.

"Reading" the Wind and the Weather
◆

Prior to each long-distance flight, human solo pilots have to make many intelligent decisions. They have to choose the best time to begin, the optimal route, and the most economical speed. When airborne, they must decide sensibly how to compensate for wind drift, how to exploit the winds, and how to choose the most appropriate altitude for different weather conditions. Unexpectedly, it was recently discovered that migrating birds make the same kinds of sensible decisions.[13]

They follow the most economical route to their destination, flying at the optimal altitude for minimizing energy expenditure while maximizing speed.

Also, they sensibly and proficiently use vertical air currents, tail winds, and other wind patterns. They take account of the weather in timing their flights and, consequently, make most of their migrations during the best weather conditions. (Their apparent ability to forecast the weather may be due to a special ability to detect changes in atmospheric pressure.[14])

A large-scale radar study of hordes of birds migrating from eastern North America to South America validated and extended these findings.[15] The migrants sensibly departed from the coast only during favorable flight conditions, and they avoided approaching storms. One set of favorable winds was used to fly over the first leg of the journey to Bermuda, and a different set of favorable winds was used to fly the rest of the way to South America. Flight altitudes varied appropriately with wind and weather. The migrants' gradually descending approach and smooth landing in South America resembled that expected of human pilots.

After reviewing these and much additional data on the migratory flight capabilities of birds, a leading expert concluded that "The adaptation of migrating birds' behaviour to hazards and hardships in the environment reaches a level of precision that is highly impressive to the most skillful and experienced human pilots."[16]

"Reading" Visual Landmarks

◆

Many kinds of birds also use landmarks as proficiently as skilled human pilots.[17] Some species, such as Canada geese, learn their migratory routes by paying close attention to and remembering the numerous successive landmarks on their first trip when they follow their parents. Some migrants use landmarks in a straightforward way; for example, they may follow coastlines or river valleys over long legs of their journey.[18] Other migrating birds skillfully compensate for wind drift and return to their selected direction by noting landmarks and using parallax, much as skilled human pilots do.[19]

Birds can discriminate landmarks much better than humans. First of all, the birds' high flight altitude provides them with a panoramic view of landmarks extending to the far horizon. Even when migrating at a relatively low altitude, say two thousand feet, a bird can perceive the landscape on a clear day stretching sixty miles in all directions.[20] A variety of avian visual capabilities supplements this expansive view. As a group, birds have large numbers of retinal cone cells which provide enhanced color vision. An individual oil droplet in each cone also serves to increase the contrast and the contour of everything seen. Birds also have a special pigment for detecting near-ultraviolet and ultraviolet light and a deep pit, or fovea, in the retina that magnifies the slightest movement. The avian visual capability is enhanced also by a richly vascularized, heavily pigmented, comblike structure, the pecten, that stands erect in the posterior chamber of the eye. Special vision-enhancing functions have been attributed to the pecten, including enhancement of perception of movement and possible service as a sextant, aiding in making navigational calculations.[21]

"Reading" the Earth's Magnetism

◆

One of the most interesting recent scientific discoveries is that a wide variety of animals are sensitive to the earth's very weak magnetism. This sensitivity is found not only in protozoa, flatworms, snails, sea slugs, and various species of amphibia and reptiles, but also in birds and humans. The surprising human sensitivity to geomagnetism was revealed in a series of experiments conducted by R. Robin Baker and his associates.[22] At a statistically significant level, the blindfolded and earmuffed subjects named the correct compass direction they were facing immediately after they had been spun round and round on a revolving chair; they also pointed correctly to their starting location immediately after completing a very circuitous bus trip. The human subjects' ability to determine compass directions appeared to be dependent, at least in part, on sensitivity to the earth's magnetism because the ability was significantly affected when the earth's magnetic field was disrupted by the sun and when experimenters used magnets to alter the geomagnetic field around the subjects. Human sensitivity to geomagnetism may be mediated by the tiny iron-containing crystals of the mineral magnetite, which were recently discovered in human brain tissues.[23] Although sensitivity to the earth's magnetic field appears to be a dormant sense among humans, who are rarely if ever called upon to use it, it plays an active role in the migrational and navigational activities of many species of birds.

The earth's magnetism serves as a navigational aid for European robins, bobolinks, ring-billed gulls, and various other species. On overcast nights, when the stars cannot be seen, nocturnal migrants, such as indigo buntings, use the earth's magnetism as a backup.[24] Various species that are able to use geomagnetism to determine directions apparently prefer to use celestial cues, when available, rather than the earth's weak magnetic field lines. As the capricious behavior of compasses indicates, the magnetic lines are not straight; instead, they swerve and ripple over the earth, vary throughout the day (due to the movements of ions around the earth in jet streams), and are easily disrupted by working generators and large chunks of metal.[25]

As I pointed out earlier, homing pigeons use the sun as a compass. However, there is also evidence that they are sensitive to geomagnetism: They tend to change direction when they fly over magnetic irregularities; and tiny fluctuations in the earth's magnetic field generated by the activity of the sun (solar flares and sunspots) affect their direction of flight.[26] Also, under heavy cloud cover, when the sun cannot be used, pigeons that are prevented from sensing the earth's magnetic field properties (by attaching tiny magnets to the birds) tend to become disoriented whereas control pigeons that can use the geomagnetic field orient normally.[27] The sensitivity of pigeons to the earth's magnetism may be mediated by the tiny iron-containing crystals of magnetite imbedded in their neck and head tissues.[28]

These data provide an understanding of the "secret" of pigeon homing. It appears that pigeons have two compasses for determining directions—the sun and geomagnetism. When pigeons are being experimentally displaced over great distances, they continually refer to one or both of these compasses (and, as we shall see, to other cues such as odors and infrasounds). They use this information *acquired during the displacing journey* to return successfully to their home loft.[29]

The nearly nine thousand existing bird species differ very widely in every capability, including their ability to use the earth's magnetism. Some species do not use the geomagnetic field lines for orientation whereas other species do. Also, different avian species may use different aspects or dimensions of geomagnetism: Some species utilize the inclination of the magnetic field lines to the vertical; other species apparently use the polarity of the magnetic field; some species use geomagnetism to select a compass direction; other species may use geomagnetism to maintain a direction after it is selected by other means.[30]

"Reading" Odors, Infrasounds, and Other Subtle Cues

◆

Pigeons also use their sense of smell to return to their home loft after experimental displacement. Apparently they pick up olfactory information during the displacing outward journey and integrate it with the windborne smells noticed earlier at the loft. These conclusions are based primarily on an extensive series of investigations by Floriano Papi and collaborators with experimental pigeons that were raised in aviaries and released at distant sites. The pigeons' ability to orient homeward was markedly reduced, as compared to control pigeons, when they were deprived of their sense of smell (by plugging the nostrils or cutting the olfactory nerve), when their aviary had been shielded from windborne olfactory stimuli, and when the direction of the odor-bearing air entering the aviary had been experimentally deflected or manipulated.[31]

Birds with good olfactory capabilities, such as the fish-eating petrels and shearwaters, also use odors to find their nests when they return home in the hours of darkness. Other birds appear to use smells to determine what they are flying over: such as a pine forest, an urban center or an ocean.[32]

In finding their way home, pigeons can also use sophisticated auditory capabilities. They can hear sounds of very low frequency (less than 0.1 hertz) that humans do not hear.[33] These infrasounds, which provide pigeons with an acoustic world unknown to humans, are generated by thunderstorms, ocean waves breaking on the shore, and wind moving over mountains. A remarkable property of these infrasounds is their ability to travel thousands of miles with very slight attenuation! Pigeons located inland appeared to sense the ultrasounds created by the tides hitting the beaches of the Atlantic three hundred miles away. Also surprising were the recent discoveries that pigeons are sensitive to changes in atmospheric pressure, to light in the near-ultraviolet range, and to polarized light patterns.[34] To what extent pigeons use these special senses in their homing feats remains to be determined.

Sensibly Deciding to Migrate
◆

Birds weigh the advantages and disadvantages of their situation and migrate when it is the sensible thing to do. When there is insufficient food, they migrate. However, when conditions are less grim, when the possibility exists that they can survive where they are, birds decide intelligently whether to migrate or not. Individual birds that find a suitable winter territory in their home range do not leave even though all the other nearby birds of their species (including their siblings) migrate long distances. Also, most birds migrate farther during the winter if there is a sudden onset of adverse conditions.[35]

Contradicting the common notion that avian migration is a more or less robotic affair, many experimental studies have shown that birds decide rationally when to migrate. Canada geese stopped their ongoing migration abruptly when displaced to milder England but not when displaced to cold Sweden. Mallards in England, which normally stay in one place, migrated when they were experimentally hatched in cold Finland. Migrating European starlings that were experimentally displaced to a favorable wintering area remained there; those displaced to an unfavorable site continued their migration.[36] In brief, migratory birds, like intelligent human beings, "weigh up the costs against the benefits and on that basis decide upon which route to take and [whether and] when to leave."[37]

Learning with Experience
◆

As they gain experience, especially during their first summer, homing and migrating birds become more proficient in using potential information to determine direction and to adjust for displacement.[38] Inexperienced pigeons need information about both the sun and the earth's magnetic field lines to return home; experienced pigeons can use one or the other.[39] When pigeons are experimentally disoriented (by manipulating the magnetic field and transporting them in total darkness), experienced pigeons, but not inexperienced ones, are able to return home.[40] When thrown off course by storms, winds, or human experimenters, young migrating birds are typically unable to change direction and presumably perish; experienced migrants, on the other hand, typically survive the displacement by changing course.[41]

Overview of Avian Navigational Abilities
◆

The recent research data on avian navigation converge on two important conclusions.

First, birds use various kinds of information to navigate without instruments. A representative species may develop an integrated compass system (from stars, sun, and/or geomagnetism) and may relate the compass directions to an ever-growing map of visual landmarks, infrasounds, smells, wind direction, and cloud movements. Depending on the demands of their particular way of life, different species use different aspects of the available information. Some migrants rely mainly on the sun, whereas others rely more on the stars, on the earth's magnetism, or on landmarks. Most birds use several kinds of information. Individual birds of the same species may use different kinds of information, and the same bird may use one channel of information at one time and switch to a different channel at another time. As avian biologist Charles Walcott recently stated, "The more we know about animal navigation the more reminiscent it is of human navigation. . . . Like ours, bird navigation involves multiple cues and is probably conditioned by where they were brought up . . . our most recent work suggests that it may well be that pigeons growing up in different places learn to favor different systems of cues."[42]

Second, complex learning plays an important role in the navigational feats of homing and migrating birds. Day migrants typically learn the daily movements of the sun and compensate for them by constantly changing the angle traveled relative to the sun. At least some nocturnal migrants learn that all stars constantly change their positions except for the North Star and its neighbors, and they use one or more stars from this stable group as a fixed point of reference.

In brief, migrating and homing birds, like human natural navigators, reach their goals by observing, learning, and integrating a variety of subtle informational cues in nature. Let us now look at humans who navigate without instruments.

Human Navigational Abilities
◆

Humans have very little natural (instinctual) navigational ability. They can learn to navigate in two ways, but each requires arduous training and painstaking practice.[43]

Professional navigators need many years of scientific training, with a focus on mathematics and physics, and much practice before they can guide a craft to a particular goal by correctly interpreting such navigational aids as sextant, chronometer, magnetic compass, log tables, charts, and nautical almanacs.

A small number of individuals living on the scattered islands of the South Pacific are able, like birds, to navigate to a particular goal over seemingly featureless seas without instruments. These Polynesian navigators are highly admired and respected in their societies because they have special knowledge, special awareness, and very useful skills that they acquired after years of intensive apprenticeship with a master navigator and years of supervised practice.

The major aim of the Polynesian navigational training is to make the apprentice "incredibly aware and keen to con every possible cue in nature."[44] These clues include, first, the rising and setting of numerous stars, which provide the equivalent of a precise compass, and the apparent movements of the sun, which provide a supplementary compass. They also learn to interpret many subtle events in nature that provide guidelines for navigation—wave patterns, dependable wind and current patterns, cloud formations (especially noting the high clouds that form over islands), color and turbidity of the ocean surface, and fog belts where cold and warm waters meet. Reliable locations of whales, dolphins, jellyfish, and particular species of fish are included in the construction of the mental map of the territory. Reefs many fathoms down are detected by sudden changes in the surface waves and color of the water. Landmarks such as shoals, islands, and atolls also play a role in guiding the natural navigator, especially at the beginning and the end of the journey. A highly trained proficiency in using cues available in nature to forecast the weather also plays a significant role. This knowledge, interpreted and integrated with many other useful facts, such as the locations relative to home of fifty to one hundred islands, enables Polynesian navigators to perform the remarkable feat of traveling without instruments to a tiny port at a particular distant place on a vast and seemingly featureless sea.[45]

The Instinctual Background
◆

As pointed out in Chapter 3, the behavior of humans, birds, and all other animals is guided by instincts or "phylogenetic adaptations"[46] — innate programs or plans of actions that are sculptured by learning experiences (to fit each individual's personal environment) and are implemented sensibly (by moment-to-moment decisions). Innate plans that are actualized intelligently (by learning and choosing) also guide the migration, navigation, and homing of birds.

A leading synthesizer of the data on bird migration, Robin R. Baker, writes that recent research "has shown convincingly that birds are born with an internal program that ... organizes the sequence and timing of molt, deposition of fat, development and regression of the gonads, and migratory restlessness."[47] Additional components of this migrational program, which is tripped off by the shortening and lengthening of days, include: endocrine secretions; an attraction toward or a tendency to move toward a particular direction at a particular time — for instance, to move southwest in autumn and northeast in spring; and an inborn feeling (or "expectation") for what the appropriate wintering area should feel like (or be like).[48] This innate avian program, which guides the young bird to migrate to an appropriate wintering area and to return in spring to its natal area, has important commonalities with various human instinctual programs, such as the pubertal programs that guide the biology and behavior of human males and human females toward the species goal of reproduction.

The human male pubertal program also has a programmed schedule of sequential components — endocrine (testosterone) secretions, growth of beard, voice deepening, rapid physical growth, enlargement of genitals and shoulders and other body parts, sperm production, heightened sexual "restlessness" and arousal, an attraction toward or a tendency to move toward a sexual partner, and a feeling (or expectation) of what an appropriate sexual partner should be like (or feel like).

The basic principles of the avian migrational instinct—a sequentially programmed series of physiological, psychological, and behavioral effects that lead via learning and choice to a particular goal—are also found in the human female. The pubertal program of one-half of *Homo sapiens* includes sequentially programmed endocrine (estrogen) secretions, ovulation, menstruation with a programmed cycle, breast enlargement, fat deposition in certain body areas, hip widening, and so on. This pubertal program can be viewed as one part of a larger female sexual reproductive program that includes subprograms for pregnancy, childbirth, and postdelivery lactation (each tripped off by particular events beginning with the implantation of the fertilized egg). Like the avian migrational program, the human female sexual program has psychological components, such as a feeling (or expectation) for what an appropriate sexual partner should feel or be like and an attraction toward or nurturing feeling toward the newborn.

Birds such as the young cuckoo that fly to their wintering grounds alone, without parents or older birds to guide them, have an instinctual tendency (or feeling) to fly in a particular direction and an instinctual feeling (or expectation) for a proper wintering area in essentially the same way pubertal humans have an instinctual feeling for and a tendency to move toward a sexual partner. However, the instinctual tendency is not sufficient to attain the species goals of avian migration or human reproduction. Like all instinctual programs, they are interblended with relevant learning, such as learning to "read" the sun and the earth's magnetism or learning to relate with potential sexual partners, and they are implemented by many moment-to-moment intelligent choices or decisions.

There is also another very important *learning* dimension to the inborn avian navigational program. Just as humans are born with an instinctual predisposition or program for focusing on, understanding, and learning to use human language, so are many kinds of birds born with a predisposition or program for focusing on, understanding, and learning to use information in their environment to travel successfully over the earth. Just as humans intelligently actualize their inherited predispositions for understanding words and for constructing grammatical sentences, so do birds intelligently implement their inborn predispositions for understanding navigationally useful characteristics of the environment. Early in life each human learns relatively rapidly and without special tutoring the phonemes, morphemes, and syntactic and semantic units of a language while also learning the rules for forming words, combining sounds, and ordering the units into sentences.[49] Similarly, various kinds of young birds learn relatively rapidly and without tutoring the principles or rules pertaining to changes in positions of the stars, movements of the sun, variations in the geomagnetic field, wind directions, directions of odors, and location of landmarks. Birds thus have a special intelligence—an inborn, instinctually based navigational intelligence—in essentially the same way humans have a special speech-language intelligence and both of these inborn intelligences are actualized by learning and integrating various kinds of complex information present in the environment.

◆

CHAPTER EIGHT

Personal Friendships Between Humans and Birds

This chapter summarizes the reports of five individuals—a retired pet clinic owner, a professor of zoology, an English lad, a behavioral scientist, and a musicologist—who formed close personal relationships with one or more birds living freely either in a human household or in the wild. These reports, and others similar to them, are never mentioned in the literature of official ornithology because they indicate that birds are much like humans, an idea that is tabooed at the highest level by the commandment against anthropomorphism. Taboos, commandments, and official charges of anthropomorphism from the scientific establishment are, of course, totally unscientific since they block the impartial search for truth. Let us leave prejudgments, preconceptions, and prejudices behind and take a close look at the characteristics and behaviors of well-observed individual birds.

A Talking Starling
◆

A young starling that had fallen from the nest was rescued by Margarete S. Corbo, a retired pet clinic owner, and raised to maturity while living freely in her home.[1] When this male starling was nearly three months old, he astonished his rescuer by repeating his name, "Arnold," in the varying tones of voice that had been used previously in addressing him. (This need not have been so astonishing. In nature, "Starlings have a large range of vocal utterances. They make whistling, clicking, and rasping sounds, and many also imitate voices and other sounds."[2]) Henceforth, the starling's rescuer and two six-year-old boys (her grandson and his friend) began deliberately repeating particular words to him. The starling then began verbalizing more English words and within a few months was saying meaningfully *in the appropriate context* such phrases as "See you soon" (only when the person addressed was to return soon), "See you, bye" (only when the person was actually leaving), "Hi" (or "Hi, hi, there"), "Kiss Arnie" (his nickname), "What are you doing?" "How are you?" "Good night," "I love you, yes I do," and "Peek-a-boo."

Although the starling would not talk to people he obviously disliked, everyone who interacted with him in a friendly way was either shocked or pleasantly surprised when he spoke meaningfully in context, especially when he began with such statements as "Hi. Good morning. C'mere. Gimme kiss." Puzzlement increased when people realized that the starling was only a few months old and was far ahead of a human child of the same age in using words meaningfully.

Subsequently, visitors also were amazed when the starling sang songs in English (such as "Mary Had a Little Lamb"), when he sang along with recorded music that he liked, when he obviously wanted to play games such as "chase," and when he clearly communicated his feelings by body language, which included the tilt of his head, the movements of his body, the look in his eyes, and the raising of his feathers.

A Very Sociable Jackdaw

◆

A baby jackdaw that had fallen from its nest was rescued by a young English lad, Julian Leek, who had developed a deep interest in reading about birds and observing them in nature.[3] With the assistance of his father, mother, and brother, Julian raised the young male bird in the Leek household, feeding him chopped-up worms and keeping him in an abandoned bird's nest until he matured. The jackdaw was raised with caring attention and with freedom to come and go as he pleased. As time went on, the family members became enamored of him and felt "pleased" and "honored" that he chose to remain with them. The young jackdaw consistently surprised the family and all who met him because he behaved the way people are supposed to behave, but not birds.

The young jackdaw was very sociable. Before he could fly, he squawked pitifully when he was left alone; later he never remained alone—he always flew to join the others. He ate breakfast at the table with the Leek family. His social nature also was manifested in his close relationship with Julian. The young bird and the young lad played hide-and-seek, bathed together, and had regular daily sessions during which the jackdaw would hop on Julian's shoulder, make little cooing sounds, and rub his bill back and forth around Julian's ear in what seemed to be "a positive frenzy of affection." The jackdaw also had a friendly relationship with Mrs. Leek, typically following her around as she did her household chores, jumping on the bed when she was trying to make it, attacking her mop, and taking the train with her to London while riding on her shoulder.

The jackdaw clearly conveyed his feelings and emotions. When the Leeks tried to lock him in a cage for the first time, he expressed his anger very clearly, and they never considered it again. It was also obvious when he was frustrated or upset; he might, for instance, tear a newspaper to shreds. By his expressions and body language the family could see how he felt and could anticipate what he was going to do. His feelings were typically positive or friendly, but, if anyone treated him disrespectfully, he nipped the person's finger just hard enough to let him or her know he deserved respect. (Although his powerful beak could inflict a wound, he never hurt anyone.)

The young jackdaw watched television and had definite program preferences, consistently observing particular programs while falling asleep during others. He also danced "wildly" to his favorite music (a particular kind of jazz), sang his own songs, and enjoyed listening to tape recordings of himself singing his songs. He was very popular with visitors and received a great deal of attention from them. He was consistently surprising and entertaining, and his obvious happiness also tended to make visitors happy.

Friendship Between a Professor and an Owl
◆

A university professor of zoology, Bernd Heinrich, rescued a very young great horned owl that had fallen from its nest and was buried in the snow following a late-spring snowstorm.[4] He nursed the young owl back to vigorous health and guided it to the point where it could survive on its own. During three summers, the male owl at times stayed in the professor's log cabin even though he spent most of his time in the surrounding woods. Despite a consensus among ornithologists that great horned owls are "fierce, defiant, and untamable, even when young,"[5] a close relationship—with caring, mutual concern, and fondness—developed between the owl and his rescuer. Professor Heinrich's field notes contain entries such as the following:

> Like an alarm clock, Bubo [the owl] wakes me at 4:34 A.M. by drumming on the window beside my ear. He joins me for breakfast, sharing some of my pancake . . . He hops onto the back of my chair, making his friendly grunts while I caress his head, and he nibbles on my

fingers endlessly. After I have had enough of the morning session of touch-and-feel, I try to write, but Bubo keeps inserting himself between me and my pencil.[6]

Bubo comes to me and hops onto my leg. For a half hour we nuzzle, tickle, and caress.[7]

He plays rough, and so do I, but eventually he tires of it and lies down on my arms. Looking at the clock I see that we have played for one and a half hours. It seemed shorter than that.[8]

When I come back to the cabin he now always comes down from his sleeping perch to play. . . . When I read he comes next to me to perch on the table so he can nibble at the pages and on my fingers. I use the opportunity to scratch his head.[9]

It is the many varied soft and hushed sounds that Bubo makes that I find most fascinating. I hear them only when I am next to him; they are his private sounds, reserved for intimacies. . . . It is these intimate details that bond friendship and promote empathy and understanding, and you learn such things from wild animals by living with them.[10]

This close relationship, with intimate play, enjoyment, caresses, and subtle nonverbal communication between a research professor and a supposedly untamable, ferocious, predatory bird, contradicts accepted notions of what is possible in human-avian interactions. The professor-owl relationship tells us that we simply do not know the true nature of the birds that surround us and we markedly underestimate their intelligence, awareness, and humanlike qualities.

Three Parakeets with "Human" Personalities

◆

A small number of behavioral scientists have succeeded in establishing mutually respectful person-to-person relations with birds living freely in their home or laboratory. In Chapter 1, I summarized Professor Irene M. Pepperberg's project with Alex, a parrot living freely in his laboratory home and interacting with the people there the way people are expected to interact but not birds. I shall now summarize the observations of another behavioral scientist, Sheryl C. Wilson, who, during her student years, closely observed three parakeets (budgerigars) that lived in her home.[11] The behavior of these parakeets also resembled that expected of people much more than it resembled that expected of birds.

Blue Bird

The male parakeet, Blue Bird, was closely observed during his entire ten-year life span. His distinguishing characteristics were a *joi-de-vivre* exuberance and skillful competence. He manifested a high level of enjoyment and skill in every aspect of his extremely active life, especially in his joyous singing, flying, playing, watching television, relating erotically with his mate, and relating to Wilson as a personal friend. His joy, competence, and intelligence appeared to be due to a series of fortunate events in his early life.

He was purchased from a family who raised uncaged birds with "loving care" in their home. Blue Bird, then five and one half weeks of age, was the most playfully active of all the young parakeets living in the large home. He was just old enough to be separated from his parents and at the optimal age for ease of adjustment and readiness to learn new things.[12]

After a few weeks of gradual introduction to the people and all parts of his new surroundings, Blue Bird was no longer afraid and was free to fly anywhere in the house and to return on his own to his open cage, which served as his home base. (He was locked in his cage for protection only at night and when he was left alone in the house.)

Blue Bird adjusted quickly and with exuberant proficiency to his surroundings apparently because he had good early experiences with people. He was welcomed with respect and positive regard by the people in his new home (Wilson and her parents) who provided him with a healthy, hazard-free environment (as described in Appendix B).

Within a few months Blue Bird astonished Wilson when he began to communicate meaningfully by using English phrases in the correct context. He began asking understandably for things he desired, such as "Open the door" (when he clearly wanted his cage door or any other door opened), "Can I have some?" (when he wanted to try whatever a person was eating), and "Take a shower" (when he wanted the faucet turned on so that he could take a shower as it dripped).

Blue Bird was not taught words formally or directly. Instead, as he interacted with Wilson, she spoke to him as if he understood, saying words slowly and in context; for instance, she would say "Open the door" when she opened his cage door. He behaved as if he wanted to learn certain words, first watching Wilson's mouth carefully and then trying to say the words himself. He made closer and closer approximations as he repeatedly heard and tried to say words and phrases.

Soon Blue Bird vocalized meaningfully in the appropriate context such words and phrases as: "How are you?" "Watcha doing?" "Where you going?" "Hello," "Good-bye," "Good morning," "Good night," and "Such a sleepy little birdie." He also understood and obeyed requests from Wilson such as "Get down," "Please go into your cage," "Give me a kiss," and "No!" and he flew to Wilson whenever she called his name. Later, when he was three years old and had a mate, Blondie, he vocalized to her meaningfully in context such phrases as "Give me a kiss," "Pretty little Blondie," and "Blondie's so nice." This basic vocabulary, which remained with Blue Bird throughout his life, was supplemented in a fluctuating way over time as he learned new phrases, used them for months or years, and then dropped them as he became interested in new things and learned the corresponding new words or phrases.

Blue Bird's exuberant enjoyment of his highly proficient daily activities can be illustrated by describing his singing, his sexual activities, his many kinds of play, and his astonishing flying abilities.

◆ *Creative singing.* Blue Bird definitely enjoyed singing songs he composed himself and also those he heard people sing directly or via recordings, radio, or television. He sang very well, and his tunes could be identified easily by people who heard him sing. He sang in a parakeet's whistle but with the same rhythm, melody, and tempo as the version he had heard. He first became excited about human music when he was about four months old and heard a song that referred to a "Mr. Bluebird" (which he apparently recognized as his name and interpreted as referring to himself). It was a joyful song that seemed to fit his personality with such words as "My, oh my, what a wonderful day! Mr. Bluebird on my shoulder . . ." At times, he flew exuberantly about when this recording was playing, calling loudly in response to his name in the song. At other times he whistled the melody and sang the words "Mr. Bluebird on my shoulder" when that line came up. Blue Bird also listened and watched when Wilson played other cheerful tunes on the piano and, at times, sang parts or bits in accompaniment.

Blue Bird also composed his own songs, which were melodic, pleasant, and cheerful. When he and his mate Blondie were resting during the day, he often "serenaded" her with melodic tunes comprised primarily of soft twitterings together with words such as "Pretty little Blondie."

◆ *Sexual activities.* When Blondie was brought to the house to join three-year-old Blue Bird, she was placed in his cage while he was in another room. When he first saw her upon flying into the room, he landed on his cage and appeared surprised as he quickly tilted his head from side to side to see her more clearly. Then he suddenly leaped from the cage (as if it were scalding hot) and began calling loudly and excitedly while flying aerobatically in loops. Finally he landed on the cage and exploded in an outpouring of cheerful-sounding chirping, squawking, and twittering directed toward Blondie as if he were excitedly talking to her. His behavior suggested "love at first sight," which is apparently as rare among parakeets as it is among humans.[13]

Blue Bird and Blondie quickly developed a sexual relationship that was playful, joyful, happy, and natural. During their lengthy foreplay, which commonly lasted an hour or so, Blue Bird sang to Blondie and talked to her in English phrases as well as in parakeet body language. They engaged in what appeared to be "love talk"—conversationlike chirping often interspersed with statements in English by Blue Bird such as "Pretty little Blondie. Give me a kiss." When he said vocally "Give me a kiss," Blue Bird meant just that because it was always followed immediately by one or two types of kisses: They bumped their beaks together one or more times or, after turning their heads sideways, they brought their open beaks together while touching and maneuvering their tongues together (resembling human openmouth or "French" kissing).

Typical foreplay between Blue Bird and Blondie began when the two birds, who had been sitting on top of an open door, moved and leaned closer together in a strikingly loving way. While Blondie listened attentively, Blue Bird sang a pleasant song. Subsequently Blue Bird said "Pretty Blondie, give me a kiss" and, immediately after kissing, he flew in a fast zigzag aerobatic display round and round Blondie. When he then landed close to her on the door top, she moved closer to him and followed him as he ran to the other end of the door. Blue Bird fed Blondie, then they kissed again and Blue Bird once more flew excitedly. Besides appearing to be having fun, they showed a high level of excitement as indicated by enlargement of the white around the eyes (caused by constriction of the pupils), erection of face and head feathers, rapid bobbing of the head, and continual jumping, chirping, squawking, and twittering. Much of this sequence was repeated several times, with variations, and it usually culminated in mating.

During copulation, Blue Bird was on top of Blondie, "hugging" her by enfolding her in one of his wings. After Blondie had turned her head 180 degrees to look at Blue Bird, they kissed and stayed in face-to-face contact during coitus. At the culmination of the copulation they showed a peak of excitement that seemed akin to an orgasm. Indeed, Blue Bird and Blondie appeared to derive much pleasure from their foreplay and copulation since they devoted much time to these activities. They engaged in foreplay and coitus almost daily and occasionally as often as two or three times per day. (Since female parakeets require a period of time in the darkness of a nest hole or nest box in order to lay eggs, and Blondie was not provided with a nest box, no offspring resulted from these innumerable matings.)

Other well-mated parakeets have been reliably observed to relate sexually in a similar astonishing way. Immanuel Birmelin writes with authority in *The New Parakeet Handbook*:

By their courtship behavior you can tell when two birds have accepted each other. . . . Parakeets make charming pairs. They often sit together, preen each other, and rub beaks. . . . Both are sitting close to each other, and the male starts singing to his beloved. . . . As he sings he moves away from the female a couple of inches, then moves closer again and taps his bill against hers several times in succession. This performance is repeated several times with growing excitement, visible in the rapid bobbing of the head. The pupils of the eyes are narrowed, and the plumage on head and throat is still puffed up. . . . If the male's wooing is successful, the female assumes the copulating position. . . . She looks almost transfixed. . . . Now the male spreads one wing over the female and executes the copulating motions.[14]

Parenthetically, the astonishing erotic excitement that has been reliably observed in well-mated parakeets also has been noted in other avian species. However, this fact has never been publicized and is known only to a small number of writers, such as Joseph Kastner, who reports in his history of bird-watching that "One bird watcher after counting a pair [of house sparrows] copulat-

ing fourteen times in succession with five-second pauses between sexual acts, accused them of being immorally oversexed
and obsessed by 'furor amatorius, the male suffering from satyr-
iasis and the female from nymphomania.' "[15]

All ornithology texts refer to the foreplay and coitus of
birds as if they are mechanical activities. For instance, birds are
said to bump (or rub) their beaks together (not kiss), rub their
cloacas together (not have sexual intercourse), and engage in
false feeding (not open-mouthed kissing). However, the data
presented by Wilson, supplemented by the reviews of Bimerlin
and Wolter,[16] show that parakeets can manifest a sexuality that
astonishingly resembles that which might be found in an exotic
human society. The sexual precocity of parakeets—their early
sexual maturity at three or four months, their readiness to mate
at any time—is just what is needed for survival of the species in
its natural niche in the interior of Australia. There the harsh
desertlike conditions make it imperative that the birds be ready
to mate while still very young and at any time throughout the
year, whenever the unpredictable rains arrive and yield the
surplus food and water needed for raising young.

◆ *Play.* Blue Bird enjoyed playing on his gym, jumping skillfully
from one trapeze to another, climbing the ladder, and pushing
the toy wheelbarrow with his beak. He also liked to slide down
the newspaper Wilson was reading and land in her lap. Often
when she was writing he would play with the pencils, papers,
and pens—biting the corners of the paper, running with a pen to
the edge of the desk, and dropping it over the edge as he watched
it fall. Both Blue Bird and Blondie enjoyed tunneling under the
bed sheets and blankets whenever Wilson was making the bed.
Showing unmistakable signs of excitement that included erection of his head and face feathers, Blue Bird intently watched
"action" programs on television, especially those with galloping
horses, stampeding cattle, or men fighting.

Blue Bird also liked to groom both Blondie and Wilson.
Using his beak and tongue, he combed Wilson's eyebrows and

hair in essentially the same way he groomed Blondie's head and neck feathers. Also, when he wanted Wilson to groom his feathers, he asked her in the same way he asked Blondie—by slightly raising his feathers and tilting his head in a particular way.

◆ *Flying with joy.* Blue Bird consistently behaved as if he enjoyed every aspect of flight—the movement, the exercise, the sensory feeling of flying, and the peak arousal that is part of a highly skilled flight performance. He was a talented flyer with great endurance, performing complicated maneuvers like a skilled stunt pilot, using dips and sharp turns to avoid obstacles while making daily rapid flights all around the house—main floor, up and down the stairs, attic, and basement. Sharp turns and skillful maneuvering were required to fly up the angular stairway, which constantly increased in elevation, and to avoid obstacles such as closed doors and floor lamps. To express joy, he had his own unique way of flying a zigzag course at a very fast speed and then shifting to dips and sharp turns in a spectacular aerobatic performance.

In brief, Blue Bird had a unique personality characterized by a high level of activity, joyful exuberance, and competence. With full involvement and enjoyment, he skillfully conducted his daily rounds—flying, singing, grooming, playing, loving, mating, eating, sleeping. He was an excellent lover for his mate and an enjoyable companion for Wilson, with whom he interrelated in the way expected of people but not of birds.

Blondie

Blondie consistently behaved as if she enjoyed listening to Blue Bird sing to her and as if she admired him and wanted to be with him all the time. She also participated in their foreplay and coitus with definite signs of enthusiasm. Although female parakeets have as much potential as male parakeets for learning to talk, Blondie did not talk; however, she seemed to understand the English phrases that were spoken to her by both Blue Bird and

Wilson. Although some female parakeets can sing well, Blondie sang poorly and only sang when Blue Bird encouraged her to sing along with him. Also, she did not share Blue Bird's excitement in flying; she flew only to get where she wanted to go.

When Blue Bird died there was no doubt that Blondie became depressed. She immediately became cranky (scolding Wilson when she tried to interact with her), did not initiate activities, did not appear to want to do anything, ate more, and slept much of the time. To alleviate her depression, a mature male parakeet, about three years old, named Lover, was purchased and brought to the house. Blondie totally rejected him — literally turned her back on him and did not look at him for many months — and, henceforth, barely tolerated him, as if she were saying "You can't replace Blue Bird." Although she never formed a relationship with Lover, Blondie eventually recovered from her depression and formed a closer relationship with Wilson. She now joined Wilson at various times; she played at Wilson's desk while Wilson studied and seemed to enjoy cuddling against her neck, sitting in her lap, and being petted by her.

Lover

Soon after Lover was rejected by Blondie, he began behaving in a surprising way, spending a considerable part of each day flying to every mirror in the house, acting as if his image in the mirror were a female bird, and enjoying sexual play with the image. Three facts indicated that he was pretending his own mirror image was a desirable female bird. First, all parakeets are perfectly aware of the striking, unmistakable facial difference between a male and female parakeet — only the male has a bright blue cere. Second, since Lover never flew into mirrors, he was aware of their reflective properties. Third, the image always reacted unnaturally — always did whatever he did.

Lover talked to the mirror image as if it were an attractive female bird, saying, for instance, "Well, hello there cutie pie. Come here and give me a kiss." (He learned these phrases in his earlier home.) He punctuated his "love talk" with the mirror image by kissing "her" (bumping his beak against the mirror) and at times by feeding "her" (regurgitating his food into "her" mouth). Other careful observers have witnessed lonely parakeets treating their reflection in mirrors as sexual objects. In *The New Parakeet Handbook*, Annette Wolter writes that "The bird's mirror image not only consoles during hours of loneliness but it also offers a chance to satisfy social needs. To be sure, the mirror evokes sexual impulses, but a healthy parakeet is sexually active even without it and simply chooses other surrogate objects for the same purpose."[17] Also in the same book, Immanuel Birmelin writes that

> Only someone who has witnessed the tender billing and cooing of a parakeet pair, the persistent efforts that precede mutual affection, and the rituals of courtship and mating can truly appreciate how desperately a single bird seeks ways that allow expression of its unfulfilled instinctual life. . . . Thousands of male parakeets desperately turn to courting shiny objects. . . . But a mirror provides no real relief, because in natural courtship behavior the action shifts back and forth between the two partners and normally leads to the act of mating.[18]

Lover soon discovered an excellent masturbatory aid that was in the house, a large stuffed toy poodle about one and a half feet high. He masturbated by crouching down on the stuffed toy dog's snout (which simulated the back of a female parakeet in contour and size), wrapping a wing around it (the way male parakeets wrap a wing around the female when mating), and moving his cloaca around and around on the poodle's button nose (which simulated the female cloaca). He always seemed happy and excited during the "masturbatory intercourse" as he called out, chirped, squawked, and behaved as if he were with a female bird. These sexual activities took up a substantial part of Lover's day; he interacted sexually with mirror images ten to twenty times a day and usually had "masturbatory intercourse" with the stuffed poodle at least once a day.

Lover was always excited about starting each day. Early every morning he rang his bell to be let out of his cage and, if there was no immediate response, he began calling "Open the door. . . . Hurry up. . . . Don't wait." He always appeared to be in a hurry to start making his rounds to the mirrors and the stuffed poodle. Although Lover barely interacted with Blondie, Wilson, or anyone else, he appeared to be happy with his vicarious sexual affairs and coitus and never seemed depressed, angry, or ill-tempered.

Although Lover did not talk to Blondie and had never before been heard to say her name, surprisingly, he expressed concern and caring for her on the day she was dying. He forsook his mirrors and stuffed poodle and spent the entire day with Blondie, walking around her on the table where she lay dying, saying caring phrases such as "Poor little Blondie. Sweet little Blondie."

Close Friendships with Many Wild Birds

◆

For eleven years musicologist Len Howard was a participant observer in the day-to-day life of the wild birds living in the orchard around her country cottage.[19] To overcome the birds' natural fear of her, she reached out to them in three ways. First, she provided numerous boxes suitable for nesting, a bird bath, and a large well-stocked food table. Then she offered special treats by hand, such as nuts and cheese. Finally she opened her cottage to the birds—opened the doors and windows, placed food and nest boxes inside, and allowed them to enter and exit freely and to live within the cottage. As a result of her outreach and her friendly attitude, the birds lost their fear of her, stopped running away from her, and began acting the way people are expected to act but not birds.

About three months after the beginning of the project, a bird entered her cottage for the first time. Howard was near the open door when, to her surprise, a blue tit flew in, hovered agitatedly in front of her face, fixed its eyes on her, and uttered cries that definitely indicated distress. Howard then noticed that the bird's mate was just outside the door and was also watching her intently. Perceiving that the birds were asking for help, she went outdoors. The couple led her to their nesting box by flying in front of her and stopping at perches along the way to turn around and see that she was following. When she reached their nesting area, she saw that their nest material had been scattered over the ground, apparently by a marauding cat. Howard responded by reforming the nest in the nest box and replacing the eggs as the pair waited close by, silently watching. Ten days later the eggs hatched. From this episode Howard came to two profound realizations: The birds had inhibited their natural fear of her and their automatic readiness to run away from her because they *understood* she was a friend to whom they could appeal for help; and wild birds can understand and communicate with people in astonishingly humanlike ways if people approach them receptively so that they lose their fear.[20]

Subsequently, Howard observed dozens of birds of various species almost daily over the course of their entire lives—as they fledged, obtained food, flew, sang, mated, raised their families, communicated, and socially related to each other. As she enlarged the scope of her increasingly intimate observations, she made three discoveries:

◆ Each bird is a unique individual with its own unmistakable personality.
◆ Conspecific birds of the same sex and age, previously thought to be exactly alike, differ vastly in intelligence, talents, likes and dislikes, and ways of behaving and interacting with each other.
◆ Behaviors that official bird experts view as stereotyped, automatic, and totally instinctual, such as building a nest, mating, incubating, and parenting, are actually flexible and variable.

As an example of Howard's intimate observations of consecutive generations of birds, I shall summarize the life histories of three great tits, a mother tit (Jane) and two of her daughters (Curley and Twist).

Jane

Jane was very competent. First, she was an exceptionally good and successful mother. Although other female great tits in the orchard rarely succeeded in raising two families in one summer, Jane did so every year. She kept her offspring in the nest longer than the other conspecific mothers before she allowed them to fly so that they had reached their maximal physical development and thus had a better chance of survival when they left the nest as fledglings.

Her first mate died from the results of an injured leg during their fourth mated year. She acquired a new mate, but he was killed by a cat when their brood of eight was still in the nest. Jane then succeeded at the difficult job of raising the nestlings alone.

Even though female great tits are not known to sing, Jane not only sang but, incredibly, was a better singer than all the male great tits Howard observed during the eleven-year project. She wrote that Jane's "song was begun in gay, ringing tones and gradually grew softer and sweeter as it lowered in pitch. A chime of bells dying away on the wind was suggested."[21]

Jane and her first mate differed from many of the other pairs of great tits in that they behaved as if they were devoted to each other (or "in love"); they were always close together, went everywhere together, entered and left Howard's cottage at the same time, and often displayed to each other in apparently sexually provocative ways.

After rearing her second brood during her sixth year, Jane seemed very tired, did not regain her strength after her molt, and died a few months afterward.

Curley

One of Jane's daughters, Curley, was markedly insecure and indecisive and strikingly different from Jane in general competence and personality.

As soon as she was able to fly, Curley accompanied her mother Jane to Howard's cottage. From the beginning Curley was not afraid of Howard and related to her on a close, personal level. Early in life Curley began to eat from Howard's hand and became very attached to her, staying near her and sleeping overnight on a railing above her bed. When Howard fed other birds, Curley hid beneath her chair. When they left she would gently nip the back of Howard's leg to get her attention. Curley also flew to Howard whenever she called her name.

Curley seemed afraid of almost all birds and avoided them. Her avoidance could be traced to a traumatic event when, as a fledgling, her crown feathers were plucked off apparently in an encounter with another bird. Her manner, which strongly suggested self-consciousness and feelings of inferiority, was related to the fact that great tits use their feet in defending themselves and she had very small feet. Her most obvious characteristic was that she kept nervously looking at and examining her feet and toes.

Despite her serious problems relating to other birds, Curley had three mates during three consecutive nesting seasons, but each of her matings was disastrous. First of all, Curley never laid any eggs. Also, from everything Howard saw and could surmise, Curley was able to avoid coition with each of her mates. Due to her extreme indecisiveness, she made her first mate defend four nest boxes against other male conspecifics seeking territory while she tried to decide in which box to build her nest. She never decided and her mate left. The following year the same indecision was repeated with a new mate who seemed to become more and more impatient and exasperated and finally left. With her third mate, Curley actually did choose a nesting box but gave the impression of "playing at nesting," since she never completed a proper nest, never laid an egg, and always dodged her mate when he wanted coition, "never allowing it." This mate also departed. At the end of the next December "poor Curley was killed by a neighbour's cat, her remains being found beside her nesting-box."[22]

Twist

Jane introduced two daughters—Curley and Twist—to Howard at the same time. Like Curley, Twist never showed any fear of Howard. Soon after she left the nest as a fledgling, Twist took naps perched on Howard's knee; she also often perched on her hand or on her shoulder. She definitely seemed to enjoy being caressed by Howard. At times when Twist was perched on her hand, Howard would gently rub her cheek over Twist's back; Twist never moved away but instead would turn up her face and look into Howard's eyes.

Howard and Twist understood each other's significant communications. It was obvious to Howard that Twist was asking for food whenever she perched on her shoulder and looked up into her face with a pleading expression. She also understood that Twist wanted only cheese and nuts, since, when offered something else, she either "chucked it across the room" or turned her back on it and again looked up at Howard with a pleading expression. Twist apparently understood Howard's request "Give me a kiss" because she always did so and only when asked. She also understood such expressions as "I'll get some cheese [or nuts]," since she instantly flew to where the items were stored with an expectant expression. As indicated by her behavior, she also understood other similar statements, such as "Have none"; at such times she immediately looked annoyed, flew to a perch opposite to Howard, and directed a fixed stare at her.

In contrast to her sister Curley who was the same age, Twist was a very good mate and mother. She succeeded each year in raising at least one brood. Whenever Howard went to visit her nest, Twist would typically see her coming from a distance and, together with her mate, would fly to Howard, perch on her shoulder or head, and eat the cheese and nuts brought for them. Although Twist spent some time indoors with Howard during each winter, she froze to death outdoors during her fourth year.[23]

Bird Individuality and Personality

From her intimate associations with numerous wild birds during eleven years, Howard insists, with the certainty of firsthand, repeatedly verified knowledge, that conspecific birds can be easily distinguished because, like humans, they each have distinct movements, postures, emotions, behaviors, and personalities. Conspecific birds of the same age and sex differ in preferences for particular foods; in propensity to be jealous, angry, fearful, happy, and to show many other feelings and emotions; in desire to play; in musical ability; and in ability to obtain food in clever ways. Since she also could observe many birds over the years sleeping in the nest boxes in her cottage, Howard could report that they differ widely in the amount of time spent sleeping and their degree of calmness and restlessness during sleep. Conspecifics also differ widely on intelligence. Particular birds that are the most clever at opening small boxes are also the most clever at solving other perceptual-motor problems and also at understanding the meanings of such statements as "Stop that," "Get off the bed," and "Come on."

Howard's unique observations showed that birds of the same species differ widely in their behavior during all stages of the life cycle. Each fledgling, for instance, copes in its own unique way with the challenges of life when it first flies away from the nest—judging where to land, deciding how to carry out the landing, finding food and shelter for the night, avoiding danger such as predatory birds and cats, and mastering the difficulties of proficient flying and landing. When chaffinch parents stopped feeding their five nestlings and encouraged them to leave the nest, four left appropriately and landed in the trees across the adjacent open area. The fifth, however, who was robust and capable of strong flight (as proved later), remained in the nest and gave several indications that it was "afraid to take the plunge"—it crouched low with "its face contorted with fear," then it sank down on the rim of the nest, misery in its attitude and on its face. . . . At last the lone fledgling stood bolt upright and with a confident expression, fluttered its wings and was off on a long flight, terminating inside a [neighbor's] bedroom. . . . this baby flew on and on, past bushes and trees into the bedroom [apparently] because it could not face alighting once it was on the wing.[24]

The lives of most birds are so fast paced and dangerously exciting that they experience the equivalent of a full human lifetime in a few years. They are "highly keyed" with faster pulses, higher temperatures, quicker sight and hearing, and their actions are often at a pace humans cannot follow with their eyes. The rapidity and richness of avian lives is illustrated by the male great tit, Baldhead, who during his first mating season had two mates simultaneously; however, he deserted the second mate and her chicks (who all died) and then raised a second brood with his first mate (who died during that winter). During his second year Baldhead and a new mate raised two broods—first their own eight chicks and then a second brood of orphaned chicks that they rescued. During his third year, Baldhead raised one brood with his mate and a second brood alone. (His mate apparently lost interest after laying the second clutch since she took off to feed and rest in the trees.) During his fourth year, Baldhead lost his territorial battle with an invading powerful male, was badly hurt, and henceforth had a lame leg. At the beginning of his fifth year, encouraged by the attention and displays of a lively new mate, he regained his lost territory by such creative battle tactics as uttering strange noises while making a series of confusing flying leaps over his rival, hanging upside down and swinging his lame leg in the air in a distracting and confusing way, and suddenly charging his rival at a phenomenal speed.

Although ornithology texts state that young males of certain species learn their songs from older male birds, this does not prepare us for the astonishing humanlike interactions that are observed during these learning sessions. Howard reports that a young male blackbird flies to the tree where an adult blackbird is singing beautifully, perches near him, and gazes at him with a listening expression. The adult stops singing. The youngster communicates to the adult bird that he wants him to sing by staring at him, showing undeniable signs of impatience, and opening and shutting his beak (which blackbirds do when they want something). When the adult begins singing again, the youngster edges closer, faces in the same direction, adopts a listening expression, fluffs out his feathers in exact imitation of the adult's pose, and then copies each slight alteration in his posture.

> The song continues for half an hour, both birds remaining almost without movement, except that [the adult] sometimes turns his head slightly while still singing and his eyes rest on the listening youngster, whose gaze is always upon the master musician. There is an air of deep content and absorption on the faces of both birds ... At last [the adult] stops singing, stretches himself and hops to the very end of the bough; the youngster copies this stretch exactly, in every detail of movement then hops slowly after him.[25]

The mating of wild birds is also at times amazingly different from the cut-and-dry affair described in the official literature of ornithology. Howard presents an example. A male blackbird stalks and chases a female blackbird up and down an open area for several weeks. Then the male suddenly bursts out with an astonishing song dramatically different from the songs associated with blackbirds. Howard testifies that it sounded "hysterical," frightened, as if it were due to some kind of torture. The male shouted this "jumbled medley of song" while he chased the female round and round in small circles on the roof of the woodshed. Howard writes:

> He was getting more and more excited as the chase grew faster; his neck was stretched out, head-feathers were

ruffled, eyes glittering and beak opened to let fling this
volley of explosive-sounding song. [The female] often
threw glances over her shoulder in return, while leading
him. . . . Suddenly she flew to a branch beside the shed
and mating took place.[26]

The same kind of explosive, hysterical "love song" (apparently
without the round-and-round dance) was also heard a few other
times immediately before blackbirds mated.

Howard also provided a striking example of birds communicating precise details to each other. After she had accidentally
lifted the top or cover of a nest box and quickly replaced it, the
surprised blue tit mother, who had been sitting on eggs in the
nestbox, quickly flew out of the box and soon returned with
the father. The two then acted as people might act if the mother
runs to the father and says something happened to the roof of the
house: On alighting, the male blue tit immediately inspected the
cover of the nest box, checked the entire box, looked inside, and
then sat on guard outside until his mate, who had gone back to
incubating the eggs, apparently was reassured.[27]

When Howard published her book, *Birds as Individuals*, in
1953, with an introduction by the noted biologist Julian Huxley,
ornithologists apparently discounted it as anthropomorphic.
Because it contradicted the official dogma that birds are mecha-
nistic carbon copies or virtual automata, it has never been
mentioned in official texts. However, much of the scientific data
gathered in recent years agrees with Howard's basic conclusion
that wild birds are much more like people than anyone (especially official scientists) had imagined. Since the scientific data
have now caught up with Howard's intensive, intimate, natural-
istic observations, each of her contentions must be taken very
seriously and tested in further research. The new generation of
scientists may revolutionize their outlook on nature as they test,
for instance, her contention that birds can communicate *detailed*
and *specific* matters to each other via body movements, body
postures, charade or pantomime, eye expressions, subtle beak

movements, loud and also muffled sounds, and possibly other exotic and subtle channels.[28]

Len Howard, Bernd Heinrich, Sheryl C. Wilson and others mentioned in earlier chapters who formed close interpersonal relationships with birds are the forerunners of a reinvigorated human race that will live in increasing harmony with birds and other living things as the message of this book spreads. Also, these same individuals, who have succeeded in befriending birds and relating to them in essentially the same way they relate to their fellow humans, are the modern counterparts of historical figures who forged intimate human-avian bonds. For example, St. Francis of Assisi "tamed many birds for pets, or rather companions—his familiar hawk at La Verna, a loving moorhen, a family of robins . . . a pheasant who followed him like a dog . . . a pet crow . . ."[29] Similarly, birds came and perched upon the hands and shoulders of St. Rose of Lima, and a certain bird with an enchanting voice would perch beside her window at sunset and the two would sing duets, first the bird singing, then Rose in her turn, and, turn by turn, for an hour or so "when the bird sang Rose said nothing, and when she sang in her turn, the bird was silent, and listened to her with a marvellous attention."[30]

◆

CHAPTER NINE

Overview of Bird Intelligence

The large number of research results that have been discussed in the preceding chapters converge on seven very important conclusions.

First, the assumption that "birds are instinctual while humans are intelligent" has been shown to be wrong; recent scientific discoveries demonstrate that both birds and humans are innately programmed to carry out their species' specialties. Each of the instinctual programs that guide people to carry out characteristic human behaviors—such as talking, bipedal walking, and mating with a human—and each of the instinctual programs that guide birds to carry out characteristically avian behaviors—such as flying, constructing a particular kind of nest, and mating with a bird of the same species—are actualized via moment-to-moment decisions requiring intelligence and new learning. The human child implements its language instinct by attending to and intelligently learning and integrating particular (human) sounds in its environment; in essentially the same way, the young bird implements its migrational-navigational instinct by focusing on and intelligently learning and integrating information provided by sun, stars, winds, geomagnetism, and other natural phenomena.

Second, the cognitive abilities of birds have been vastly underestimated. Like humans, birds are able to generalize, form abstractions and nonverbal concepts, and discriminate and integrate many sources of information. In general, they remember, or "store" past information as proficiently as humans. At times, when it is useful in their niche, they manifest a long-term memory that is superior to that of most humans.

Third, birds communicate, or "speak," to their flockmates by calls, songs, and a complex body language that includes many subtle changes in the eyes, crest, beak, feathers, wings, and every other aspect of the avian body. They communicate to their flockmates everything that is relevant or of interest to them. The more deeply research probes into bird communication, the more certain it becomes that birds "speak" with their bodies and with their voices to each other about all the significant immediate facts of their life. Like nonliterate humans living in an outdoor environment, they "speak" to each other about the important daily events—the rain, the animals, food, water, mating, offspring, and so on. Birds such as parrots, parakeets, and starlings, which have the physical capability to articulate the sounds used by humans, can speak to people meaningfully. To what extent they learn to speak meaningfully depends primarily on how much patient effort humans devote to relating with them, bonding with them, speaking intentionally and contextually with them, and teaching them words and phrases. The evidence indicates that birds can come to understand the meaning of human vocalizations and body language just as humans can come to understand avian vocalizations and body language.

Fourth, we now know that the assumption "birds are instinctually driven machines," led earlier investigators to miss the flexibility in avian behavior. The accumulated evidence has now made this flexibility obvious. It can be seen in birds who sensibly and flexibly vary virtually every aspect of their behavior in their winter and summer homes, for instance, behaving as solitary, day-active, insect eaters in their northern home and as social, twilight-active, fruit eaters in their southern abode. Avian flexibility also is seen in the drastic changes in behavior with variations in the food supply—not laying eggs or not incubating the eggs when there is insufficient food (avian "birth control"), raising more offspring when there is plenty of food, and sensibly not feeding the weakest chick in the brood when there is insufficient food for the parents and all the nestlings. Avian flexibility is also manifest in the way conspecific birds build their nests differently depending on the nature of the local predators.

Fifth, in their natural environment, birds enjoy playing in various ways that resemble the play of human children. Perhaps someday all human children will have the opportunity to enjoy playing together with birds. Also, since each bird is as much a unique individual as any human, each new bird our children get to know well can be as interesting as a new friend from a very different culture; the next generation of humans can have the same kind of exciting pleasure with birds that was experienced by field anthropologists a century ago who first befriended an African pygmy, an Australian aborigine, a Hopi, or a Samoan.

Sixth, birds experience the same fundamental feelings and emotions as humans—they can be contented and happy and even ecstatic as well as sad and hopeless and forlorn, and they can manifest parental love, close friendships, and erotic love. They can show concern and benevolent caring for a mate, for their offspring, for their siblings, and for ailing or crippled flockmates and even, at times, for members of other species (including *Homo sapiens*). (Befriended birds can not only be friends to our children but also share their concerns and show affection and love for them.)

Seventh, birds have a sense of the beautiful, a general aesthetic sense, just as humans do. They can also compose and sing songs that human musicians admire and can sing duets, trios, quartets, and even quintets antiphonally or polyphonically.

Avian Superiorities

◆

Birds have several sets of capabilities that are superior to those of educated humans in technological societies (but not necessarily superior to those of nonliterate humans in tribal or nonindustrial societies). These superior capabilities include, for example, the following.

Birds are able to earn a living and to raise a family outdoors without special aids or tools. They can procure from the natural environment a variety of nest-building materials, sufficient water, and also food items of just the right kind and right size to feed themselves and their voracious and fast-growing nestlings. They are able to build a safe and secure home, exerting much sensible effort over a considerable period of time to find just the right materials and interlace them together in a tight-fitting way. In building their homes, birds can manifest the skills of a tailor, mason, or other human craftsman without the need for special tools.

Using information found in the natural environment, migrating and homing birds can determine both precise direction and precise passage of time (the "avian compass" and the "avian clock"). They can use natural information to "read" barometric pressure, wind patterns, the earth's magnetism, polarized light patterns, subtle odors, movements of the sun, patterns and movements of the stars, infrasounds, and subtle landmarks. They can use these natural cues to find their way much better than either modern humans or most nonliterate tribal humans.

As a group, birds are more alert than people, and they react faster to dangers approaching from above, below, or the sides. Their tempo of living is speeded up or accelerated. Most species of birds race through life at a much faster pace than humans and they crowd many more events into each bit of time—growing, metabolizing, maturing, and mating and reproducing faster (sometimes within a few months after birth), and dying much sooner. Because of this temporal speedup, most kinds of birds have as many experiences in their brief lives as humans have in their much longer but slower-paced lives.[1]

Human Superiorities

◆

Although birds are superior to humans in various ways, humans also have their superior capabilities, specifically tool making and language.

Birds can make simple tools when they are useful in a particular niche. However, with their highly developed, manipulative hands, their superior ability to reflect, and their splendid brain-eye-hand coordination, "primitive" or tribal humans can make far more intricately complex tools than can birds. Moreover, combining their highly developed tool-making ability with their equally highly developed ability to reflect and symbolize, "sophisticated" modern humans have created the advanced tools of present-day technology that far surpass the capabilities of birds.

Tribal and nonliterate humans have a complex language that uses spoken words as symbols to represent or "stand for" ideas, events, and objects. These spoken symbols can communicate abstract notions, subtle ideas, and information quickly and proficiently from one person to another. Although birds communicate their intentions and ideas by sounds and body language, the avian language appears to be much more concrete and less abstract than the languages of tribal humans. The symbolizing specialty has developed to an extreme in technological societies. Modern humans use many kinds of symbols—spoken words, hand signs of the deaf, written symbols, Morse code, braille, television images, computer images, and so on—to stand for and to transmit proficiently from generation to generation a huge number of abstract ideas. This transmission of symbolic information down the generations is the basis of human culture. Its many symbol-based institutions include religion, art, education, law, and socioeconomic-political structures. As anthropologists have been emphasizing for nearly a century, humans differ from other species primarily in that they *record* thoughts in books, records, carvings, folktales, papers, and the like and pass them down to future generations as accumulated cultural knowledge. Since other animals do not carry recorded knowledge across generations, their past thinking is lost and people think they are thoughtless.[2]

CHAPTER TEN

Why Birds Have Been Totally Misunderstood

Why has mankind misjudged the basic nature of birds? Why does virtually everyone mistakenly believe that birds are much more like machines than like people? How can our most advanced thinkers ask so naively, "Are we the only intelligent life form in the universe?" when the birds they do not notice nesting outside their windows are perfectly aware of them and are behaving intelligently in the demanding natural world? There are many surprising reasons for this total misunderstanding. Let us look at them with an open mind.

Ignorance of Birds as Individuals

◆

Very few people ever become personally acquainted with one individual natural bird. They simply never interact with a bird in nature because practically all wild birds are afraid of people and thus do not let them approach too closely. Rare exceptions, seen primarily in the 1700s, were birds on isolated, unexplored Pacific islands that did not flee from humans when they first approached. Since all wild birds that share an environment with humans now fly away immediately upon their approach, they must be aware of the general reality that humans can hurt them, and they may be aware of the specific reality that humans kill them, eat them, and think nothing of them.

After careful thought, Julian Huxley, the great biologist, emphasized that "Only when birds have come to lose their fear, can a human observer really begin to be let into the secrets of their lives, and discover the degree of their intelligence. This point [is] to be taken to heart by professional biologists."[1] How many avian biologists, ornithologists, ethologists, and comparative psychologists have taken Huxley's admonition to heart? How many have reported on the life experiences and behaviors of one particular bird that was not afraid of humans? Virtually none. (The exceptions were discussed in Chapters 1, 5, and 8.) Up to this time, avian science has provided vast accumulations of data on the behaviors and habits of bird species but virtually no data on the life of one single individual bird. Whenever avian scientists describe the behavior of a particular bird, they perceive it as a member of a species, not as an individual living its life over time. This misperception derives from the mistaken assumptions that birds resemble programmed robots, that the behavior of conspecific birds of the same sex and age does not differ significantly, and that "if you've seen one, you've seen them all."

Scientists are not the only ones who are totally ignorant of individual birds. There are many reasons why practically all "civilized" people believe birds are nothing worth thinking about individually. One important reason is that "Science says so." Since science influences present-day opinions in essentially the same way religion used to, nonscientists assume that "birds are nothing" because they know, as members of modern culture, that official science says birds are like machines.

This brings us to one of the great paradoxes of modern

science. The scientific data on birds deal with birds as members of the class Aves or as members of particular species; it provides useful knowledge about the characteristics of birds in general and the instincts and habits of particular species. But with a few notable exceptions, such as the research with Alex the parrot, the "hard" scientific data do not help us understand a single individual bird with its unique personality and life experiences. It is as if intelligent beings from another planet studied earth's humans and drew valid conclusions about them in general—for instance, how humans in general procure their food by hunting, fishing, gathering wild plants, herding livestock, and agriculture or how they kill each other in wars—but never say anything about the personality or unique experiences of one single person.

With very few exceptions (see S. C. Wilson's report in Chapter 8), personal relations between humans and their domesticated birds are also either nonexistent or, at best, extremely superficial. There are two distinct categories of domesticated birds: chickens and turkeys raised for human food and pet birds kept in cages or enclosed aviaries. Let us look first at how humans relate to their domesticated poultry and then at how they perceive their caged pet birds.

Domesticated Birds

◆ _Poultry._ Chickens and turkeys raised on farms for their meat and eggs are not considered individuals for several reasons. First of all, their owners and handlers are strongly motivated to perceive them as entities without a significant degree of awareness. Since they have to kill the birds and sell them for meat, it would be extremely bothersome to see them as having feelings or as being conscious. Second, poultrymen "know" from everything they have been told by everyone, including scientists, that birds are robotic. Third, domesticated chickens and turkeys are "crippled" birds. For thousands of years, humans have controlled their food supply, territorial rights, mating, and every other aspect of their lives, and they have been bred for fast growth, large size, meaty breasts, lightweight bones, manageability, and minimal intelligence. Consequently, they have lost a significant part of their avian nature and are markedly different from red jungle fowl and wild turkeys (their wild counterparts). Our long-enslaved or "domesticated" chickens and turkeys do not act at all like natural birds; they're virtually desexualized, they cannot live without the assistance of humans, and they soon die out when returned to nature.[2]

Even though chickens and turkeys have been deprived of their survival capabilities and a significant part of their instinctual intelligence, they still possess surprising mental capabilities. For instance, they perform at an unexpectedly high level of intelligence (equal to that of nonprimate mammals) on formal "learning-to-learn" tests in which they must first learn a general problem-solving principle and then use it to solve similar problems in different forms.[3] Also, in every large flock, each hen knows all the others as individuals and has an agreed-upon pecking order or ranking as inferior or superior to every other hen. Although this ranking is usually determined by actual combat, it is also at times determined simply by the mere threat of combat, with the more aggressive and confident bird as the victor. The status hierarchy serves a very useful purpose; constant fighting within the flock is prevented because each chicken knows when it can take its turn at the food trough and the dusting area and knows its place in the roost. (Roosters also have their own status hierarchies, which likewise are determined by individual combat, and high-status males mate more often than their rivals.)[4]

◆ *Caged pets.* People also fail to see the individual personalities of the pet birds they keep in cages. The bird that is naturally meant to be free is confined to life imprisonment by a giant being who is typically about twenty-four times as tall and has about one thousand times the bird's mass. The person who keeps the bird permanently imprisoned in a cage assumes, along with everybody else, that the bird is a stimulus-response machine that does not have real feelings (such as wanting to be free). An odd but prevalent assumption is that the bird *belongs* in its cage, the way a pet fish belongs in its aquarium, because, in the words of one sophisticated cage-bird owner, "it is a domesticated cage bird, not a wild bird, and it is comfortable in its cage and adapted to its niche." No one asks the obvious: To what extent would I be understood by a giant being who kept me imprisoned and thought of me as a machinelike being without feelings or awareness that belongs in a cage? (Before allowing your pet bird to fly freely, see Appendix B to learn how to maintain an uncaged bird in your home with minimal hazards.)

The Fallacy of Size
◆

Most humans have an ingrained belief that tiny things just can't be intelligent and aware. Although they might consider a mammal near their size or larger—a chimpanzee, a dolphin, a whale, an elephant—as having some degree of intelligent awareness, they find it very difficult to imagine that creatures as small as songbirds could be intelligently aware. The accepted assumption, that a certain minimum size is necessary for intelligence, is fallacious. If the proper relations are maintained among the parts of an organism, a reduction in its size or its miniaturization does not affect its intelligence or awareness. The smallest recorded adult human, who was roughly one and one-half feet tall, lived with as much intelligent awareness as the tallest person, who towered around ten feet. There are no indications that the tiny dragonflies of today with wing spans of about two inches function less well than the giant dragonflies of the Carboniferous period with wing spans exceeding two feet. Also, comparing across classes or phyla, the tiniest mammals, the shrews, which are often smaller than songbirds, are at least as intelligent as the largest cartilaginous fish, the sharks; some insects are smaller than some protozoa; some vertebrates are smaller than some insects; and some invertebrates, such as the giant squid, are much larger than almost all vertebrates.[5] To what extent very small animals or very large ones are or are not conscious and intelligent should obviously be determined by the evidence, not by a priori assumptions or beliefs.

The Small-Brain Fallacy
◆

The misunderstanding of birds is also due to the pervasive "small-brain fallacy" and to an associated "small-cortex fallacy." The latter fallacy is based on a syllogism:

The cerebral cortex is the seat of intelligence.

Birds have very little cerebral cortex.

Therefore, birds have very little intelligence.

The fallacy in this logic is that the cerebral cortex is the seat of some of the human specialized intelligences, but not necessarily of other kinds of intelligence. Humans have developed the cerebral cortex to carry out such specializations as hand-tool manipulation and symbolizing. Birds have developed a different part of the brain, the hyperstriatum, to carry out their specialties, such as navigating without instruments. The hyperstriatum is just as hypertrophied in birds as the cerebral cortex is in humans.[6]

The brains of certain birds, such as the well-studied canary, have an astonishing capability that is lacking in the human brain. These bird brains are able to enlarge their size, *when needed*, to carry out a particular activity. It was recently discovered that, when male canaries compose their repertoire of entirely new songs each spring, there is a tremendous enlargement of the "highest vocal center," that part of their brain responsible for composing the songs and also for controlling the muscles that vocalize them. In autumn, when both song composition and singing go into abeyance until the next spring, the hypertrophied higher vocal center shrinks to about half its size; the next spring it nearly doubles in size again.[7]

In brief, the conventional arguments that bird brains are too small or do not have particular structures needed for intelligence are based on ignorance of brains in general and bird brains in particular. Furthermore, even if the avian hyperstriatum was not hypertrophied and the avian brain could not enlarge when needed, it would still be unwarranted to argue that the small brains and small bodies of birds render them less capable of behaving with intelligent awareness than animals with large brains and large bodies. The argument is as unwarranted as contending that the small government and small size of Liechtenstein render it less capable of behaving with intelligent awareness in the international arena than a large nation such as China with a huge government.

Misunderstanding of Instincts

◆

Misunderstanding the nature of birds is linked to misunderstanding the nature of instincts. As documented in Chapter 3, data accumulated during the last three decades have clearly shown that the conceptions of birds as instinctual and humans as intelligent is fallacious. Let us look more closely at the instincts and intelligence of both birds and humans.

All animals, including humans, have a large number of instincts needed to survive and reproduce in their niche and to carry out all of the activities that characterize their species. Birds have an instinct to fly, to mate in the avian way, and to communicate by species-specific calls and body language. Humans have an instinct to walk, mate in the human way, and communicate by human vocal language and body language. All animals, including humans, implement their instincts by moment-to-moment intelligent judgments that fine-fit the instincts to the complex and changing environment. Although birds have an instinct to fly and humans have an instinct to walk, they both implement and sharpen their instincts by moment-to-moment decisions and long-term practice. (There is, however, a quasi-exception to the fine-fitting of instincts. In all animals, including humans, the instinct to obtain nourishment has to function proficiently at birth, with negligible practice. Virtually immediately after birth, precocial birds are able to peck for their food, altricial birds are able to gape to solicit food, and human infants are able to suckle, even though all three ways of obtaining nourishment become more proficient with practice.)

The nature of instincts was most clearly revealed by the pivotal discovery that mankind's vaunted linguistic-symbolic intelligence is instinctual. Ironically, this instinctual specialized intelligence underlies the speaking, reading, and writing of humans and all their cultural institutions—their science, religion, philosophy, mathematics, art, literature, socioeconomic-political structures, and so on. Virtually everything that is special to humans, that differentiates them from other animals, is a product of their specialized instinctual intelligences—their linguistic-symbolic intelligence and their hand-tool intelligence. It is of deep significance that virtually everything that is special in any species, that differentiates it from other species, is also due to its instinctual intelligences. Consider the beaver. It differs from all other animals in its instinctual hydroengineering intelligence, which underlies its ability to change its environment for its own benefit by building well-designed dams. It uses the dammed-up water to build canals to its food sources, to provide safe storage spaces for food beneath the ice during the winter, and to provide a solid home for the beaver family rendered doubly safe from predators by the water surrounding it on all sides.

Young humans implement their language instinct by attending to, learning, practicing, and using the words and grammar they hear. In essentially the same way, young birds implement their navigational instincts by intelligently focusing on, learning, practicing, and using the information that is present in nature. Also, the human mother who is nurturing her newborn is as unaware of the instinctual basis of her behavior as the avian mother who is feeding her newly hatched chicks with gaping mouths. These considerations, and others presented in Chapter 3, converge on the unexpected conclusion that humans are just as instinctual and just as intelligent as their neighbors, the birds, and that both taxa have their own unique special intelligences.

Misunderstanding Intelligence

◆

Humans believe that their specialized linguistic-symbolic intelligence and tool-manipulating intelligence are the components of intelligence per se. They fail to realize that there are many other kinds of intelligence, some of which are found in *both* humans and birds, including musical, social, bodily-kinesthetic, and spatial intelligence.[8] Humans also are aware that they have practical intelligence—the intelligence needed to discriminate causes and to use these discriminations to bring about desired effects. However, they fail to realize that birds also perceive cause-effect relations and thus also have practical intelligence. In brief, there are different kinds of intelligence; some are found in both humans and birds, others are unique specialties of humans, while others are unique specialties of birds.

Human Narcissism

◆

People in modern technological cultures generally assume that humanity is supreme not only in all dimensions of intelligence but also in every positive quality, such as the ability to feel with the feelings of others, to feel joy, and to mate erotically. The data presented in this book, however, sum up to the conclusion that humans are the best in symbol-using and tool-making specialized intelligences (which have wide ramifications in literate human cultures and in modern technology) but are not necessarily superior in any of the other kinds of intelligence or in any of the positive qualities such as empathy, joy, and eroticism. The deep narcissism of humans is a powerful barrier to their accepting a paradoxical conclusion: Birds may feel superior to humans because they may correctly perceive that they "own" the air; they swiftly go where they please; they literally look down on the humans below them; they use the information available in nature to migrate and navigate from one rich food source to another; and, with no need for tools, they procure their sustenance, raise their offspring, and build secure homes in every part of the earth including many habitats inhospitable to humans.

Human Benefit

◆

A defining characteristic of Western technological culture is its highly efficient destructive exploitation of the earth and its nonhuman species for the benefit of humans. This unusual relationship between humans and other species, which did not exist prior to the industrial revolution, is an entirely different thing from the natural predator-prey and hunter-hunted roles in nature that have maintained ecological balance on the earth for hundreds of millions of years. This unusual exploitive relationship between humans and birds and humans and all other animals is another very important reason why birds have been totally misunderstood. Mankind's exploitation of birds is rationalized and justified by the notion that birds are instinctual machines and do not really matter. This powerful motivation to relieve self-criticism and guilt by categorizing birds as stupid and inferior has unfortunate consequences for mankind. It prevents humanity from understanding the natural world and from living enjoyably in mutual harmony with the many-splendored living creatures that comprise earth's biosphere. In Chapter 12, which discusses the implications of intelligence and awareness outside of *Homo sapiens*, we shall see why this destructively exploitive relationship with nature has to change quickly to a respectful

relationship if the human species (and many other species) is to
remain alive and healthy.

Anthropomorphic Dread
◆

Modern humans also have been blinded to the fact of avian
intelligence and personality by the dogma of official science that
it is anthropomorphic (and thus supposedly unscientific) to say
that birds or other animals have human characteristics. The
dogmatic commandment "Thou shalt not anthropomorphize!"
has a surprisingly interesting and little-known history. It did not
become part of official science until some time after the great
Charles Darwin and his close associate, George Romanes, had
published their influential writings on the emotions and intelligence of birds and other animals. In two now-forgotten books
published a dozen years after his world-shaking *Origin of Species*,[9]
Darwin concluded from a wealth of evidence that despite the
specialized intelligence of the human race (manifested in its
tool-making ability and its ability to reason), differences in
emotions and mentality between humans and the other animals
are differences of degree and not of kind. Also using the data
available at that time (primarily observational reports rated as
reliable from naturalists and educated Englishmen), Romanes
documented that birds and other animals act intelligently in a
wide variety of circumstances in much the same way one might
expect from "simple folk."[10]

After the books by Darwin and Romanes were published,
an attack began on both men, especially on the younger and less
famous Romanes, accusing them of gullibility. They were
charged with this sin because, it was said, they not only uncritically accepted fallible personal testimony as evidence but they
also used this "anecdotal" evidence to anthropomorphize or
humanize animals—that is, to describe animals as if they are
people when they obviously are not. This debate between the
"soft-headed" Darwin and Romanes and the new generation of

Birds of different species can show a caring concern for one another, and at times form a pair bond (Chapter 1). Here a red rosella tenderly grooms its cockatiel companion.
Hans Reinhard/Bruce Coleman Ltd.

Above: A woodpecker finch has selected an appropriately sized cactus spine to use as a tool in flushing out wood-boring insects (Chapter 1).
Alan Root/Bruce Coleman Inc.

Below: This heron is fishing with bait (bits of food strategically placed in the water). It watches intently and is ready to catch fish when they bite (Chapter 1).
Larry Lipsky/Bruce Coleman Inc.

Above: Birds defend their nests in flexible ways (Chapter 2). This red-winged blackbird first tried to deter the swan by hovering over it and now directly attacks it, since it continues swimming towards his nest.
Erwin and Peggy Bauer/ Bruce Coleman Ltd.

Right: A male weaverbird skillfully applies the methods of basket weavers and of weavers using looms to construct an individually designed, secure home for fragile eggs and restless nestlings. Developing this level of skill takes practice (Chapter 2).
Cyril C. Laubscher/Bruce Coleman Inc.

Above: Learning to sing (Chapters 6, 8). A young wood thrush listens to a singing adult whose song he will later attempt to copy.
Jack Dermid/Bruce Coleman Ltd.

Below: Courtship among birds can include impressive male performances (Chapter 8). Here a male cockatoo gives an upside-down display of his acrobatic skill.
Alan Root/Bruce Coleman Ltd.

A satin bowerbird demonstrates a keen esthetic sense as he decorates the dance floor of his courtship bower by artistically adding flowers, shiny objects, and colorful feathers while pausing after each addition to study the effect (Chapter 6).

Michael Morcombe/Bruce Coleman Inc.

Above: Birds enjoy playing. This adult secretary bird is playing alone as it tosses a clump of grass in the air and then jumps to catch it (Chapters 6, 8).
Jane Burton/Bruce Coleman Inc.

Below: After losing a mate, some birds have demonstrated reactions that resemble human grief (Chapters 8, 10). Here a pheasant coucal is apparently trying to revive its mate, killed by a car.
Brian Coates/Bruce Coleman Ltd.

Above: Bird courtship often seems joyful (Chapter 8). Here a pair of grebes run in unison across the surface of the water in one phase of their impressive courtship dance.
M.P. Kahl/Bruce Coleman Inc.

These mated budgerigars (parakeets) illustrate the neglected fact that avian couples, much like human lovers, may caress, cuddle, kiss, and carry out other loving interactions that may or may not be followed by the act of mating (Chapter 8).

Hans Reinhard/Bruce Coleman Inc.

"hard-headed" scientists went far beyond disagreements about interpretations of data, and reflected primarily the contrasting philosophies of its two sets of antagonists. To understand the nature of the data that were attacked as "anecdotal," "anthropomorphic," and "unscientific," and to determine their validity in light of all the evidence now available, let us look at four representative reports from Romanes's chapter on the intelligence of birds.

Reporting on a pair of mated mandarin ducks living in his large aviary, an English gentleman testified that, when the male drake was taken away, the female duck "displayed the strongest marks of despair at her bereavement, retiring into a corner, and altogether neglecting food and drink, as well as the care of her person. In this condition she was courted by [another] drake who had lost his mate, but who met with no encouragement from the widow." When her original mate was subsequently recovered and returned to the aviary, "the most extravagant demonstrations of joy were displayed by the fond couple; but this was not all, for, as if informed by his spouse of the gallant proposals made to her shortly before his arrival, the drake attacked the luckless bird who would have supplanted him, beat out his eyes, and inflicted so many injuries as to cause his death."[11]

A writer on biological topics shot a tern in a flock and went to retrieve it. Then, he wrote:

> I beheld, to my utter astonishment and surprise, two of the unwounded terns take hold of their disabled comrade, one at each wing. . . . After being carried about six or seven yards, he was let gently down again, when he was taken up in a similar manner by the two who had been hitherto inactive. In this way they continued to carry him alternately, until they had conveyed him to a rock at a considerable distance, upon which they landed him in safety.[12]

Dr. Franklin, a "distinguished observer," reported in the scientific journal *Zoologist* that a male parrot took care of his sick mate by carrying food to her in his beak and aiding her feeble attempts to raise herself and that, when she was dying,

> Her unhappy spouse moved around her incessantly, his attention and tender cares redoubled. He even tried to open her beak to give some nourishment. . . . At intervals he uttered the most plaintive cries; then, with his eyes fixed on her, kept a mournful silence. At length his companion breathed her last; from that moment he pined away, and died in the course of a few weeks.[13]

Three independent reports from reliable observers showed that birds are capable of soliciting the assistance of conspecifics to obtain revenge on birds of a different species that have expropriated their nest.[14] Typical of the three reports is the one by an English naturalist who wrote that, after a pair of swallows failed to remove a sparrow that had taken over their nest, they took flight and returned shortly afterward with a number of other swallows, "each of them having a piece of dirt in its bill. By this means they succeeded in stopping up the hole, and the intruder was immured in total darkness. Soon afterwards the nest was taken down and exhibited to several persons, with the dead sparrow in it." The naturalist commented that "In this case there appears to have been not only a reasoning faculty, but the birds must have been possessed of the power of communicating their resentment and their wishes to their friends, without whose aid they could not thus have avenged the injury they had sustained."[15]

Soon after the death of Darwin and the early death of Romanes, a new hard-headed, positivistic Zeitgeist, or spirit of the times, swept through Western thought and especially through animal psychology. This no-nonsense, antianecdotal, experimental approach was championed around 1900 by such powerhouses as C. Lloyd Morgan, Jacques Loeb, and Edward Thorndike, carried forward by John Watson through the early 1920s, and subsequently enshrined as the unquestionable orthodoxy by three subsequent generations of behaviorists.[16] This approach maintained that a solid science cannot be constructed on anecdotal data. Instead, a science of human and animal behavior must be built on a bedrock of reliable data from controlled experiments. Over time this behavioristic, experimental approach became increasingly sophisticated methodologically with an emphasis on six procedures: random assignment to experimental treatments; quantification, precise measurements, and utilization of probability statistics; replication of experiments; experimental controls including controls for subtle effects such as the unintended effects of the experimenter on the experiment; proficient use of increasingly sophisticated equipment (from mazes, puzzle boxes, and polygraphs to the most advanced electronic technology); and parsimonious and reductive interpretation of results with the focus on stimuli and responses and "mechanisms."[17] This methodology had a virtual stranglehold on research in comparative psychology until about 1970 and remains the dominant paradigm even today.

The behavioristic-mechanistic-stimulus-response paradigm has stimulated a large number of controlled experiments and careful observational studies, many of which are summarized in previous chapters, for instance, Pepperberg's studies with Alex the parrot, Hernnstein's experiments with laboratory pigeons, and the research on learning and memory in birds. The behavioristic experimental research that has now accumulated shows, ironically, that Darwin and Romanes were basically correct in their conclusion that humans and birds (and also other animals) are not essentially different in their basic feelings and practical intelligence. Since both the earlier "soft" research and the later "hard" research discovered avian intelligence, it is time for official science to adopt a genuinely scientific (knowledge-seeking) attitude toward the topic of animal intelligence and look with an open mind at the data when scientists report that the birds they carefully observed behaved like people rather than like machines.

◆

CHAPTER ELEVEN

Are All Animals Intelligent?

It is time to turn to the revolutionary im-
plications of the discovery spelled out in this book that birds, like
humans, are not machines but intelligent, willful individuals with
distinct personalities. First of all, this new scientific discovery
implies that official science and consequently Western culture
have up to now misconstrued the nature of reality. Since official
science has misperceived our closest neighbors, the birds, it must
humbly start again step by step to move toward a truer conception of the natural world. First, it is crucially important to
determine if the intelligent awareness that characterizes humans
and birds is also present in other animals. There are at least three
major possibilities: (1) humans and birds are the only really
intelligently aware animals; (2) humans and birds are the two
animals with the highest degree of intelligent awareness in a
continuum that includes all animals; or (3) all animals within
their own niches are equally aware and intelligent.

How humans are to understand themselves and the world
around them and how they are to live their lives now depends on
the answer to this pivotal question: Is willful intelligent aware-

ness present in all animals, in some animals, or only in humans and birds? Surprisingly, the answer to this question can already be discerned in the voluminous research that has accumulated during the past two or so decades on the behavior of numerous animals other than *Aves* and *Homo sapiens*. Despite the realistic fear that they will be charged with the sin of anthropomorphism and excommunicated from the church of official science, many courageous scientists have dared to report, albeit in very muffled tones, that the animals they carefully observed living freely in nature or in human homes behaved essentially like unsophisticated humans — with awareness, practical intelligence, flexibility, willfulness, intentionality, foresight, and mindfulness.

In this chapter we will review astonishing research conducted with four representative kinds of animals — apes, cetaceans, fish, and hymenoptera — which changes our conception of reality. We will see that intelligence may be common throughout the animal kingdom. In the last chapter we will look at the revolutionary implications of pervasive animal intelligence.

Intelligent Apes
◆

The accumulated research on apes converges on the conclusion that these animals are conscious, sensitively aware individuals with practical intelligence and distinctive, humanlike personalities. We will look first at two long-term projects with a gorilla and a chimpanzee who were raised by humans and who behave shockingly like unsophisticated people. Although neither gorillas nor chimpanzees have the physical capability to articulate the sounds used in human speech, they can learn American Sign Language and can hold signed conversations with humans. Apes that have learned to use sign language are telling us that they are conscious and aware, they understand everything significant in their situation, and they understand people as well as people understand them.

Koko, the Gorilla

Koko, a female gorilla, was born in the San Francisco Zoo. When she was about a year old, she began to learn American Sign Language, which is as complete, intricate, communicative, and grammatical a language as English, German, or Chinese. It includes many hand signs, each one representing or symbolizing a concept and functioning in the same way as a spoken word. For instance, slapping the thigh means "dog." It is used by the deaf in the United States to converse with each other as proficiently as hearing individuals converse in spoken English.[1]

Koko's teacher was Francine (Penny) Patterson, who at the beginning of the project was a graduate student at Stanford University working on her doctoral dissertation.[2] After two years of visiting with and working with Koko in the zoo, Dr. Patterson succeeded in moving Koko first to a mobile home on the Stanford campus and then to a small farm. Koko, currently age twenty, now resides there in a mobile home next to Patterson's house.

Koko learned to use American Sign Language surprisingly quickly. There were two kinds of instructions: As Koko listened and observed, Patterson (or one of her assistants) spoke a word and simultaneously made its sign by appropriate hand movements; also, Koko's hands were typically guided or "molded" to help her make the sign for the word being taught.[3] Astonishingly, within one month, one-year-old Koko was communicating by correctly using the signs for *food, drink, more, gimme, come, that, out, up, dog,* and *toothbrush.* After two months of exposure to hand signs, baby Koko was signing two-word combinations or "little sentences," such as *Gimme food, More food,* and *Drink there.* By the third month she was asking questions. After three years of exposure to American Sign Language she had a working vocabulary of more than two hundred words; after ten years of exposure she was appropriately using more than five hundred signs. This number has continued to increase.

The fact that Koko learned to use a very large number of

the signs she was exposed to is less illuminating than her simple but meaningful conversations. Let us look at representative examples of her statements while keeping in mind an important point: When translated literally into English, American Sign Language is bound to seem telegraphic and ungrammatical because it has a different grammar from spoken English and a unique economy of expression. For instance, it uses neither the verb "to be" nor the "little words"—the articles, conjunctions, and prepositions such as "the," "and," and "to."

Koko is asked to pick the gorilla skeleton from pictures of four animal skeletons. She does so correctly.

PATTERSON (SIGNS AND SAYS): Is the gorilla alive or dead?
KOKO (SIGNS): *Dead, good-bye.*
PATTERSON: How do gorillas feel when they die—happy, sad, afraid?
KOKO: *Sleep.*
PATTERSON: Where do gorillas go when they die?
KOKO: *Comfortable hole bye.*
PATTERSON: When do gorillas die?
KOKO: *Trouble, old.*

The research assistant did not leave at her accustomed time.

KOKO (SIGNS): *Time bye you.*
ASSISTANT: What?
KOKO: *Time bye good-bye.*

PATTERSON (SIGNS AND SAYS): What did you do yesterday?
KOKO (SIGNS): *Wrong, wrong.*
PATTERSON: What wrong?
KOKO: *Bite.*
PATTERSON: Why bite?
KOKO: *Because mad.*

PATTERSON: Why mad?
KOKO: *Don't know.*

ASSISTANT (SIGNS AND SAYS): What do you say when you really want
    to insult people?
KOKO: *Dirty.*
ASSISTANT: Okay, can you think of another one?
KOKO: *Sorry, gorilla polite.*
ASSISTANT: It's okay to tell me.
KOKO: *Toilet.*

When Patterson wrongly scolded her, Koko signed, *You dirty bad
toilet.* When an alarm clock finally stopped ringing, Koko commented in sign language, *Listen quiet.* When given a toy bird with
a white breast, Koko signed, *Bird have stomach white.* When
touching a velvet hat, Koko signed, *That soft.* When punished by
confinement, Koko has signed, *You open door out now, I be good, be
quiet* and also *Out me please key open.* When asked to say a long
sentence about lunch, Koko signed, *Love lunch eat taste it meat.*
When a visitor arrived, Koko signed, *Happy good you come.* Koko
also signs to herself when she does not know she is being
observed; for instance, she picks up one of her blankets, smells it,
and signs *That stink.*

    By creatively combining signs she knows, Koko invents
meaningful signs for objects whose sign she does not know. For
instance, not having learned a sign for wedding ring, she sensibly
called it *finger bracelet.* With the same kind of creative inventive-
ness, Koko spontaneously referred to a zebra as a *white tiger,* to
tweezers as *pick face,* to a cigarette lighter as *bottle match,* to a face
mask as *eye hat,* to a rich tapioca pudding as *milk candy,* to a lemon
(whose taste she did not like) as *dirty orange,* to a pineapple as
potato apple fruit, and to stale cake as *cookie rock.* She also created
her own iconic or pictorial sign for thermometer by tucking the
index finger under her arm where a thermometer had been
placed previously. Similarly, she created her own original and

very understandable iconic signs for stethoscope, nail file, and tickle.

With some adaptations to the standardized instruments, it was possible to assess Koko using tests originally designed to measure the intelligence of children, such as the Wechsler Preschool and Primary Scale of Intelligence and the Peabody Picture Vocabulary Test. Her estimated intelligence quotient, or I.Q., which ranged from roughly 70 to 90, was within the low range for American children of the same age. Surprisingly, she performed better than an average American child of the same age in detecting what was missing in incomplete drawings and also in carrying logical progressions to completion. As might be expected, she performed poorer than an average American child in fitting puzzle pieces together and also on tasks that called upon the human specialty of fine motor control, such as penciling a path through a maze or making detailed drawings. Examples of Koko's answers to items on the intelligence tests include:

Name two animals? *Cow, gorilla.*

What lives in water? *Tadpole.*

We drink out of a cup and also out of a ? *Straw.*

Milk and water are both good to ? *Drink.*

What can you think of that's hard? *Rock* [and] *work.*

Koko's "free associations" and definitions also demonstrates her awareness and intelligence: What do you think of when I say Koko? *Me.* Shoe? *Foot.* Wash? *Good clean.* Drink? *Sip.* Orange? *Food drink.* Michael [a male gorilla friend of Koko's]? *Toilet devil.* Thanksgiving? *Nice good food.* Smart? *Know.* Sad? *Frown.* Wrong? *Fake.* What is the opposite of first? *Last.* What do you do with a stove? *Cook with.* What does it mean to be tired? Koko signs *Sleep* and mimics a yawn. Are you happy with who you are? *Koko love good.*

When shown a photograph of her birthday party, Koko signed, *Me love happy Koko there.* When a research assistant who had drilled Koko on the names of the parts of the body (eye, ear, nose, and so on) asked Koko what she thought was boring, Koko signed, *Think eye ear eye nose boring*. When Koko developed a "crush" on a young man and eagerly observed the gate through which he was due to arrive, she signed *There look eager*.

When a research assistant showed a white towel to Koko and asked its color, Koko signed *red*. Assistant: You know better, Koko, what color is it? Koko signed *red* three more times. Then, with a grin, Koko picked up a minute speck of red lint that had been clinging to the towel, held it up to the assistant's face, and signed *red* again. Koko apparently enjoyed this "trick" or "joke" since she repeated it later with Patterson.

Koko behaves as if she understands exactly what is said to her. For instance, when Patterson signs and states, "After you kiss your baby [doll], I'll give you these leaves," Koko hunts for the doll, finds it on the other side of the room, kisses it, and returns to Patterson with hands outstretched for the leaves.

Even though Koko never received formal instruction in spoken English, she understands it and is able to translate from spoken English to American Sign Language. For instance, several toy animals were arranged in front of Koko and she was asked: Which animal rhymes with hat? Koko signs *Cat*. Which rhymes with big? Koko points to the pig and signs *Pig there*. Which rhymes with hair? Koko points to the bear and signs *That*. Which rhymes with goose? Koko points to the moose and signs *Think that*.

Koko's understanding of everything that matters in her immediate environment, and her practical intelligence and awareness, is also demonstrated by the fact that over the years she has assisted in cleaning her mobile home; cooperated with various chores; brushed her teeth; groomed her fingernails; looked at pictures in magazines; played with her toys, dolls, and precious live pet kitten; deliberately dismantled her toilet; played with a younger male gorilla by wrestling, chasing, hiding, tickling, and mutual grooming; and let out a cry of grief and was inconsolable for days when she was told her beloved kitten had been killed by a car.

Humans have long been certain that they alone have intelligence and awareness, they alone understand the meaningful things and events around them. This human self-centered conceit, which is leading to devastation of the natural world, is basically false. Koko, the gorilla, has clearly shown that she intelligently comprehends everything that affects her in her human-controlled environment. Even though Koko's feats have been generally ignored by official science or mistakenly brushed aside as due to unconscious experimenter bias (commonly referred to as a Clever Hans effect),[4] it is crystal clear to unbiased and nonintimidated scientists that Koko has the basic characteristics of a sensible person.

Washoe, the Chimpanzee

The same kind of basic intelligence that underlies the signed statements of Koko, the gorilla, is also seen in chimpanzees who have learned American Sign Language. Consider the female chimpanzee Washoe, who converses in American Sign Language essentially the same way as Koko.

Washoe was raised from ages one to five by Dr. Allen Gardner and Dr. Beatrix Gardner at the University of Nevada.[5] Soon after Washoe learned her first eight signs, she spontaneously began combining them; by the time she reached age five (when the formal research project was discontinued), she said (via sign language) hundreds of little sentences, such as *More food, Hug me good, Listen dog* [barking], *Come open, Time eat? Washoe sorry, You tickle me, Please give Washoe sweet-drink, You me go out hurry,* and *George smell Roger* [who was smoking]. While Roger Fouts (who was a major participant in the project) smoked and ignored her requests, Washoe signed five consecutive statements: *Give me smoke, Smoke Washoe, Hurry give smoke, You give me smoke, Please give me that hot smoke.* While the research assistant, Susan, continued stepping on Washoe's rubber doll, Washoe signed fourteen consecutive statements: *Up Susan, Susan up, Mine please up, Gimme baby, Please shoe, More mine, Up please, Please up, More up, Baby down, Shoe up, Baby up, Please more up, You up.*

Representative examples of Washoe's "little conversations" in sign language include the following:

At the end of lunch.

WASHOE: *Enough.*
PERSON (SIGNS): Enough what?
WASHOE: *Enough eat time.*

Washoe's doll is nearby.

PERSON (SIGNS): That my baby.
WASHOE: *No, no mine.*

Before going outdoors.

WASHOE: *Out out.*
PERSON (SIGNS): Who out?
WASHOE: *Washoe want out.*

Washoe knew early on that a sign did not refer just to a particular object but abstractly to all the objects that shared its characteristics—the sign for baby referred to babies of various species, the sign for dog meant all the types of dogs and also their pictures. She understood the sign for "open" to refer to opening doors, boxes, drawers, jars, briefcases, and magazines. She correctly used the sign for "more" to refer to an abstraction covering many particular kinds of increase in quantity, and she generalized the sign for "dirty" so that it became a generalized insult or condemnation—*Dirty leash, Dirty cat,* and, when Roger displeased her, *Dirty Roger.* She also understood the crucial change in meaning that occurs when the order of words in "Roger tickle Washoe" is changed to "Washoe tickle Roger." Also, when she did not know the sign for an object, she could creatively invent a sign; for instance, she called a watermelon a *candy drink*, a swan a *water bird*, and a Brazil nut a *rock berry*.

Toward the end of the formal project, when Washoe was about five years of age, she used signs throughout the day, starting conversations, making comments, asking questions, and signing or "talking" to herself (especially when playing with her dolls or looking at pictures in magazines). She used signs to refer to many (and possibly all) objects and events in her environment that were significant for her. These included signs for everything edible or drinkable, all significant inanimate objects, all living creatures in her environment, the personal names of all her human acquaintances, verbs referring to all actions or events that were significant to her, and many other words including five colors, generic names, pronouns, possessives, locatives, comparatives, qualities, and materials. Furthermore, in a childlike way, Washoe ate at the table with forks and spoons, cleared the table and washed the dishes, dressed and undressed herself, used the toilet, played with children's toys, liked picture books and magazines, and acquired a credible level of skill with screwdrivers and hammers.

More Symbol-using Chimpanzees

Two young male chimpanzees participating in a project conducted by Sue Savage-Rumbaugh also showed a surprising ability to use symbols to communicate meaningfully. Following a training period, the chimpanzees, Sherman and Austin, used a special set of symbols to communicate about things not in the immediate situation. In this way they demonstrated that, like humans, they understood the symbolic function of symbols.[6]

In another project by the same investigator, a young pygmy chimpanzee (*Pan paniscus*) named Kanzi learned to understand the English spoken in his environment when no attempt was made to tutor him. He learned the meanings of words and phrases in essentially the same way a human child learns them, by attending to spoken words and observing the context in which they are used. At age four Kanzi was tested for English comprehension in a formal experiment that rigorously excluded experimenter bias effects.[7] By carrying out the actions requested, Kanzi demonstrated that he had spontaneously learned to understand the exact meaning of numerous statements, such as: *Kanzi tickle* [a particular person or animal] *with a stick* [or other implement]. *Take the green beans* [or carrots or other objects] *to* [a particular location, such as the group room or the campfire]. *Put the rock on the vacuum.* Put the onions [or tomatoes or other objects] *on the fire. Go get* [one of many objects] *that's in the bedroom. Plug in the vacuum. Chase Bill* [or another particular person] *with the stick* [or another implement]. In her formal summary of the results with Kanzi, Savage-Rumbaugh concluded that he comprehends spoken English, identifies the visual symbols that the researchers have assigned to stand for spoken words, comprehends the symbolic use of symbols, and uses symbols when their referents are absent. Equally important, he acquired these capabilities in essentially the same way they are acquired by human children, not by conditioning but by observation.[8]

Socially Sophisticated Chimpanzees

Wild apes living in nature also show the social awareness and individuality seen in Koko and Washoe. Two dozen chimpanzees living freely in a large open enclosure in Arnhem, Holland, carefully observed for several years by Frans de Waal, behaved like sociable but distinctively individualistic people who are members of a somewhat exotic primitive culture. Some female chimpanzees were much more interested in sex than others; some were much more maternal; some were much more popular with the males. Some of the males were much smarter than others; for instance, all of the researchers and attendants at Arnhem agreed that one young male chimpanzee, Dandy, not only consistently outwitted the others but also organized all of their coordinated escapes and ingeniously maneuvered his caretaker to bring him daily large, satisfying quantities of bananas, his favorite food. Also, the researchers were astonished to discover that the chimpanzees are "adept at the subtle political manoeuvre. Their social life is full of take-overs, dominance networks, power-struggles, alliances, divide-and-rule strategies, coalitions, arbitrations, collective leadership, privileges, and bargaining."[9]

During her long-term project observing wild chimpanzees, Jane Goodall noted many unexpected humanlike behaviors: The chimpanzees engaged in social hunting and social eating of small animals such as monkeys; when their territory did not provide sufficient food, they carried out bloody warfare with a neighboring chimpanzee colony to expand their territory; they constructed and used simple tools, such as skillfully made termite mound probes; they proficiently communicated everything they needed to communicate to each other by body language and special vocalizations; and they had "tremendous fun" with "sheer exuberance" during their special parties with "a good deal of noise and much play, mating and grooming . . . a carnival atmosphere."[10]

Intelligent Cetaceans
◆

Recent research discovered startling humanlike behaviors in dolphins and an unsuspected specialized intelligence in whales.

Dolphins

In a superbly conducted project, two bottlenosed dolphins (*Tursiops truncatus*) were taught words presented either as whistled sounds or as hand gestures. After they learned the words, the dolphins demonstrated they understood requests, such as the following, by doing precisely what was requested: *Go over* [or under or through] *a hoop. Put the basket in the net. Take the surfboard to the Frisbee. Put the net on the basket. Take the Frisbee to the surfboard.*

Fetch the ball [or hoop or Frisbee] *to a* [named] *dolphin. Touch a pipe* [or ball, hoop, fish, person, surfboard, basket, net, Frisbee] *with your fluke* [or pectoral fin], *grasp it with your mouth.* The two dolphins learned about thirty words that could be arranged into about 1,500 to 2,000 uniquely meaningful sentences that they understood![11]

In another recent project, pairs of dolphins understood what "do it together" means and complied by carrying out (simultaneously by mutual coordination) such behaviors as leaping together high into the air while continuously touching snouts ("kissing") or continuously touching flippers ("holding hands").[12] In another set of replicated experiments, rough-toothed dolphins (*Steno bredanensis*) understood the meaning of "do something new, different, or creative" by carrying out a bizarre series of creatively new behaviors never before seen in the species, such as skidding along on the floor of the tank; keeping only the tail continuously out of the water; and in one continuous action swimming in a series of figure eights, jumping out of the water, skidding across six feet of wet pavement, and tapping the experimenter's ankle with the snout.[13]

In their natural environment, teams of dolphins show their willful intelligence as they maneuver strategically as a group to catch large fish and also to repel predatory sharks. In fact, as a science writer noted in reviewing the recent research, dolphins resemble humans in their complex

> social system featuring lasting friendships, altruism, babysitting, cooperative defense and elaborate communication. They also seem to enjoy the very human pastime of sex for sex's sake. . . . Dolphin mothers are highly cooperative. A group of them will often form a "playpen" around their calves so that the young can interact in a protective enclave. "Babysitting" and "aunting," in which one female watches another's calf, are also common. . . . Dolphins live in an altruistic society.[14]

Although research during recent decades has demonstrated that dolphins are aware and intelligent in the same basic way as humans, this striking discovery is minimized by the researchers in the area. Apparently they are afraid they may be accused of the venal sin of anthropomorphism. A recent overview of his own research by a leading investigator, Louis M. Herman, highlights both the intelligence of dolphins and its minimization in matter-of-fact, unemotional "scientific" language:

> The dolphin readily forms and applies rules about relationships . . . rules that generalize to the whole class of problems. . . . Perhaps the most impressive auditory accomplishment was the ability of a dolphin to learn to understand sentences . . . understanding was shown for novel instructions as well as for more familiar ones. . . . Both the semantic and the syntactic features of sentences were taken into account . . . she [the dolphin] was able to match geometric forms or abstract forms easily. . . . A major theoretical result of this study is that the dolphin was able to understand references to absent objects as well as to objects that were present. A practical result is that one can interrogate a dolphin about the contents of its physical world and receive accurate reports of those contents.[15]

This research stimulated a science writer to exclaim emotionally that it

> will force humankind to reconsider its role in the world vis-à-vis animals. It may force humans to look at themselves as an extension of animals, and to look at animals as extensions of themselves. It may be an incredibly humbling experience for humankind; it may demonstrate that there has been all along more intelligence in nature and its systems than in all the collected wisdom of human beings.[16]

Whales

A surprising musical intelligence, resembling the musical intelligence of humans and birds, has also been discovered in humpback whales (*Megaptera novaeangliae*).[17] During each breeding season about 90 percent of the male whales appear to be trying to sing the same song—one that is "fashionable" or "popular" at that time. However, each male sings this song in his own way and, as a group, they are continually changing it so that, after a few singing seasons, it has been changed completely and they are now trying to sing a different "fashionable" song. (The songs of the other 10 percent of the males are considered "aberrant" because they deviate significantly from the popular song.)

Since the sounds of whales travel far in the water without attenuation, the humpbacks listen to each other's songs over distances of hundreds of miles. To keep up with the continually changing popular song, the whales must notice small changes in the themes, phrases, subphrases, and notes. Although the songs of humpback whales in the Pacific differ from those in the Atlantic, all humpback male songs have striking structural and aesthetic similarities. All of the songs are eerily beautiful. The foremost researcher on humpback whale song, Roger Payne, described it as a "vast and joyous chorus of sounds . . . sounds that boomed, echoed, swelled, and vanished as they wove together like strands in some entangled web of glorious sound . . . sheer ebullience . . . lovely, dancing, yodeling cries."[18] The songs have musical structure. They are comprised of four to ten themes sung in the same order, and each theme is a unique set of musical note sequences—phrases and subphrases. A repeated rhymelike pattern embedded in the songs apparently serves as a mnemonic device and is used by the whales to recall the phrases and subphrases of the elaborate compositions.[19] Of vast significance for understanding musical intelligence is that, when played at high speed, whale songs are indistinguishable from bird songs; at an intermediate speed, they can be mistaken for possible human compositions. Apparently, birds, humans, and whales possess a basic musical intelligence since they can listen to, appreciate, create, and sing intricate and beautiful music that is executed by each taxon at a different tempo.

Intelligent Fish
◆

Everyone knows that a fish does not have a mind. We all learned, very early in life, not to think of fish as aware, conscious, intelligent beings. This deeply ingrained belief has been shown to be erroneous by recent thorough research with three kinds of fish—cichlids, "cleaners," and gobies.

From intensive studies of cichlid fish, two experts concluded that they "are intelligent fishes. . . . They betray some of those attributes which make man the successful animal that he is. They exhibit an awareness of their surroundings, a definite alertness, and what we might inadequately describe as inquisitiveness."[20] Also, cichlid fish deceptively pretend they are dead until a smaller prey fish approaches close enough to snap up and eat; they burrow in the sand to avoid fishermen's nets; they communicate their feelings, intentions, and concerns via an elaborate language that involves changes in body color together with emission of particular odors and controlled sounds; and they shepherd their young and alert them to danger.[21]

Several dozen scientists have reported that various kinds of fish congregate at particular places in the sea and interact in an intelligent, socialized, mutually profitable way.[22] For instance, predatory fish that wish to be cleaned travel to particular cleaning stations where "cleaner fish" are waiting to do the job. The little cleaner fish benefit by removing and eating the parasites and other foreign material on the skin of the large predatory fish. The predators also apparently like to have their teeth cleaned— they allow the cleaner fish to enter their mouths and to eat all the food particles stuck between their teeth. Of utmost importance is the surprising discovery that when the predatory fish meet the cleaner fish anywhere in the ocean except at the cleaning stations, they immediately snap up and eat them. Many species of cleaner fish (at least fifty are known) and of predatory fish come together at the cleaning stations, which are located at particular places in the sea, such as at old shipwrecks or beside a coral colony.

The little cleaners and the large predators intelligently cooperate for mutual benefit; the predators take their place in line, wait their turn, stand still patiently while being cleaned, and, at the right time, open their mouths and uncover their gills so that the cleaner fish can enter and safely clean these areas. During a six-hour period of close observation, roughly three hundred fish were cleaned at one cleaning station in the Bahamas.[23] Since researchers are aware that they will be ridiculed and expelled from official science if they mention the striking similarities between these cooperating fish species and cooperating human groups, they say nothing about the obvious humanlike aspects and mention only "scientifically acceptable" implications—writing, for instance, "From the standpoint of the philosophy of biology, the extent of cleaning behavior in the ocean emphasizes the role of co-operation in nature."[24]

Another series of experiments showed astonishing humanlike intelligence in goby fish (*Bathygobius soporator*). These fish live where the ocean meets the shore. At low tide they are caught in tidal pools. However, they are able to jump from one pool to another with amazing accuracy—they practically always land in the next pool and do not become stranded on the dry land. Even though there are often numerous separate pools in the immediate environment of these goby fish, they have been observed leaping without error from one to the other until they have traversed them all. Three thorough experimental investigations showed that the gobies cannot see the next pool before leaping, are not using other sources of direct sensory information about the location of neighboring pools, and are not using trial-and-error learning of the jumps.[25] These results, together with the crucial observation that the gobies will not jump when placed in an unfamiliar pool, led to the startling conclusion that, when swimming through their environment at high tide, these fish learn and remember the detailed topography of the ocean floor, including each depression and elevation, and then, using this knowledge, deduce or calculate how far to jump in order to leap into the next pool. Each component of this feat—attending to, learning, and remembering innumerable useful details about the environment; planning ahead before carrying out a useful activity; deducing or calculating the distance to the next pool; and coordinating the leap so as to land in the chosen spot—are components of a special kind of intelligence that has eerie resemblances to the kind of intelligence that humans consider (erroneously) to be unique to *Homo sapiens.*

These recent findings indicating intelligent awareness in various fish species are harmonious with some forgotten (or discounted) scientific reports. The earlier reports stated that various kinds of fish watch over their young and convey them to a new location when danger threatens, show marked increases in caution only when appropriate (for instance, when swimming through much-fished waters), and manifest unmistakable feelings such as curiosity when examining an unfamiliar object, anger when a male's territory is invaded, and jealousy when fighting over particular females.[26] Apparently, humans have misperceived not only the intelligent nature of birds, apes, and cetaceans but also of fish and, as we shall see in a moment, of other animals. It is time for humans to realize they do not understand the basic nature of reality, to put aside their conceited arrogance, and to look again and reperceive more truly everything around them. They can begin now to look with new eyes at the previously unnoticed, stepped-upon ant, to which we now turn.

Intelligent Hymenoptera

◆

Ants

Official science has assumed that ants are tiny machines prewired to carry out inflexible behaviors. This assumption has been proven wrong by hard-headed research demonstrating that ants are socially cooperating individuals acting sensibly in their lilliputian niche.

After many years of carefully observing the way ants responded to each other's sounds and body motions, the leading myrmecologist of the past generation, William Morton Wheeler, was startled to realize that ants "actually communicate with each other."[27] The two leading current myrmecologists, Edward O. Wilson and Bert Hölldobler, similarly conclude that ants use many modes of communication including "tappings, stridulations, strokings, graspings, nudgings, antennations, tastings, and puffings and streakings of chemicals."[28] Some ant messages can be easily translated by humans. For example, in one carefully studied species, "Let's go to a particular area" is communicated by a special kind of body jerking together with movements of the antenna and an odor trail. "Let's go fight the invaders" is conveyed in essentially the same way except that the body jerking is much more intense.[29]

Although symbolic communication is highly developed in humans, it is not unique to them; it is also seen in ants. For instance, the message "Let's go fight the invaders" is passed on in a chain, each ant informing others about an event (the invasion) that has not been observed but has been passed down symbolically by body language.[30] Humans can easily miss both the subtle sound language of ants (such as their barely perceptible stridulating sounds, which call other ants to their rescue) and also their subtle body language of barely perceptible movements of the jaws, antennae, abdomen, and other body parts.

Their communications by odors (pheromones) have been elegantly elucidated in recent years. Only a small number of odoriferous molecules are needed to carry the message from one ant to another. Ants do not react to the odors in a rigid or mechanical way. Instead, amazingly, they *interpret* the *meaning* of each odor differently depending on its quantity and also on whether it is emitted alone or as part of a blend of odors.[31] A particular emitted odor will be interpreted in one context as meaning "Flee" and in another context as meaning "Fight." Also, different specialists or castes respond differently to the same message. For example, when informed that their nest is being invaded, the soldiers typically rush toward the intruders, some of the workers grab pupae and carry them deeper into the nest, the remaining workers join the soldiers in the battle, and the reproductive female stays in her chamber but assumes the most effective protective posture.[32]

The important discovery that ants actually communicate — that is, transmit meaningful, interpretable information to each other essentially like people — should not have been surprising. After all, myrmecologists had long known that the ability to communicate meaningfully was necessary to explain the coordination of the ant colony's many cooperative activities. The integrated, harmonious functioning of the ant colony is due to the cooperation of the individual ants not only in foraging, defending the nest, caring for the eggs and nurturing the young (the larvae and pupae), but also in less obvious tasks such as constructing coordinated nest corridors in three dimensions, keeping the interior of the nest clean, maintaining special toilet areas, and removing the dead.[33]

To comply with the dominant reductionistic-behavioristic *Zeitgeist* and to avoid the charge of anthropomorphism, present-day researchers try to interpret ant behavior as inflexible and mechanical. Nevertheless, shining through the research data is the flexibility and variability of ants, which appear to be associated with an ability to select or choose among alternatives. Examples of the nonpublicized discoveries that demonstrate ant flexibility, variability, and choice include the following.

It was recently discovered in four harvester ant species that the females apparently *select* the males with whom they mate.[34]

Defenders vary their behavior to deal flexibly with nest invaders. For instance, a defender may stand perfectly still until one invader approaches closely enough to be crushed suddenly with the mandibles, but the same defender may unexpectedly approach another invader from the rear and attack it at particularly vulnerable parts.[35]

It is well known that some worker ants are consistently speedy and others sluggish.[36] A leading entomologist, Rémy

Chauvin, has dared emphasize "the greatest importance" of the discovery that two conspecific worker ants of the same age often respond to the same situation very differently. For example, when caterpillars that were known to be killed and eaten by an ant species were offered individually to workers of that species, only 25 percent of the ants attacked and continued fighting the caterpillar until it was dead; 33 percent fought it but not to the bitter end; 25 percent did not attack it; and 17 percent retreated from it as if afraid.[37] Edward O. Wilson writes that in at least some ant species, " 'personality' differences are strongly marked even within single castes."[38] In experimental escape tests, ants differ widely on a bell-shaped normal curve with extremes of very fast and very slow escapees.[39]

The individuality and flexibility of ants is also evident when experimenters move them to a new or different situation. When foragers are transferred experimentally from outside the nest to the area inside the nest where the larvae and pupae are located, they flexibly take on the job that is now appropriate—they begin to take care of the young. Also, closely observed foragers differed widely in the percentage of time they socialized, foraged, and slept or rested. Even when busy, ant societies do not differ from human societies in the proportion of time devoted to labor and idleness.[40]

Like humans, ants can alter their environment artificially to reach their goals. They have been reliably observed tunneling under obstacles, bridging a large span by holding on to each other in a particular way, and constructing bridges by placing tiny pieces of earth bit by bit across a liquid barrier.[41] Their learning abilities are also surprising. In the laboratory, ants have learned difficult mazes with as many as ten blind alleys and have remembered them when tested four days later.[42] Like many types of birds, many kinds of ants learn to use the sun as a compass by associating its apparent movements across the sky with geographic directions and the passage of time. In fact, ants of many species can return to their nest by the shortest direct route no matter how circuitous the outward journey. They accomplish this astonishing feat apparently by learning to use their sun compass to measure the angle (direction) of each leg of their outward journey; keeping track of the distance traveled along each leg; and, finally, computing the shortest distance between the end point of their journey and the point where their nest is located.[43]

Different colonies of the same ant species can behave very differently. In one carefully studied species, members of one colony lived on dead insects, members of another colony "stealthily" robbed ants of other species, another colony lived on sprouting corn kernels, and another lived on the food bits in a human kitchen.[44]

There is no truth in the official science dogma that ants are inflexible in their basic activities, such as in building their nests.

If a colony's preferred nest site is unavailable, it can adopt a strikingly different type of nest site, such as cow dung. In fact, some species build their nest in virtually anything that can be excavated.[45]

When a colony is excavating a powdery, dry soil, some workers flexibly and sensibly change their behavior and travel to a distant source to obtain water to wet the soil.[46]

The ant colony flexibly abandons its nest and moves in an organized way to a new nest site whenever it is the intelligent thing to do; for instance, ant colonies have been shown to move when the nest is disturbed by an experimenter, when there is insufficient sunshine due to overgrowth of tree branches, or when there is a restriction of foraging space or hunting grounds by a competing ant colony.[47]

Some colonies have a winter nest in a protected area and a less secure summer nest where food is abundant. At scattered places along their major routes, ants may construct a series of very small nests or retreats used by foragers to rest or to escape pelting rain or extreme midday heat.[48]

When an experimenter deliberately damages a nest, the ants reconstruct it according to a "systematic order of work, with such plain marks of an intelligent plan, and carried forward much after the manner of men . . . with practical wisdom, and with ready adaptation of means to the new condition."[49]

Humans and chimpanzees are not the only animals that carry out bloody warfare. During the past two centuries a large number of myrmecologists have described their careful observations of wars—fierce physical combats with many dead—between different ant species and also between different colonies of the same species. This vast literature has been elegantly summarized by a thoughtful scholar:

> Every kind of warfare known to ourselves will be found in the world of ants; open warfare, overwhelming assaults, levies *en masse*, wars of ambush and surprise and surreptitious infiltration, implacable wars of extermination, incoherent and nerveless campaigns, sieges and investments as wisely ordered as our own, magnificent defenses, furious assaults, desperate sorties, bewildered retreats, strategic withdrawals, and sometimes, though very rarely, brawls between allies, and so forth.[50]

However, like humans, ants do not always settle their territorial disputes by bloody warfare. Again, going back two centuries, there are numerous careful reports of ants settling their disputes by elaborate one-on-one tournaments in which very few are injured. Some of these tournaments resemble human wrestling matches; others resemble human boxing matches. An early report of such a tournament between two species in territorial conflict states that the ants "reared up on their hind legs, two by two, and wrestled with one another, seizing a mandible, a leg, or antennae . . . overturning one another, falling and scrambling up again . . . on every hand there were groups of ants struggling together, and I never saw any of them emerge from the combat wounded or mutilated." A very recent report on an ant tournament by the two foremost present-day myrmecologists states that

> When two hostile workers meet, they initially turn to confront each other head-on. Subsequently, they engage

in a more prolonged lateral display, during which they raise the gaster even higher and bend it toward the opponent. Simultaneously they drum intensively with their antennae on and around each other's abdomen, and frequently kick their legs against the opponent . . . each ant seems to push sideways as if she were trying to dislodge the other. After several seconds one of the ants usually yields and the encounter ends. . . . They soon meet other opponents and the whole procedure is repeated. . . . If a large and a small ant are matched in a displaying encounter, usually the smaller one yields. . . . In fact the behavioral analysis of the displaying patterns suggests that during encounters the contestants gauge each other's size, and that there is a tendency among the ants to bluff, that is, pretend to be larger than they really are.[51]

Humans are not the only animals that have the intelligence to live by farming or by keeping cattle or slaves. Ants also possess these abilities. Consider these reports.

Some kinds of ants make their living by cooperative agriculture with a fine division of labor. For instance, 190 different species of ants raise a nutritious fungus crop within their nests by forming work groups that cooperatively carry out many well-coordinated activities: Specialists bring in fresh leaves from the outside; the leaves are processed into compost within the nest by work teams that cut the leaves into tiny pieces, crush the pieces into a pulp, and enrich the pulp with saliva and droppings before allowing it to go through a process of controlled decay; other teams weed out unwanted fungi species; and still other teams specialize in controlling the ventilation, temperature, and humidity by constructing just the right number of entrances, which they then open and close at the appropriate time.[52]

Many kinds of ants keep aphids in the way various human societies keep cattle: They "milk" their aphids to obtain their sugary liquid honeydew; they may build "barns" for their aphids, and they always protect them from predators; and they sacrifice and eat a few aphids when they need protein instead of sugar.[53]

A substantial number of ant species procure slave ants by conquering another colony via expert maneuvering and a coordinated battle plan.[54] Observing these surprisingly effective maneuvers during ant slave raids, a careful observer, A. Forel, reported that "a human army robbing a foreign town or fortress, could not behave better or more prudently."[55]

The research just reviewed leads us to the same conclusion as the earlier research reviewed by Romanes and Darwin. Romanes concluded: "Many of the foregoing facts display an astonishing degree of intelligence . . . among ants. . . . Some at least of these foregoing facts can only be reconciled with the view that the insects know what they are doing and why they are doing it."[56]

As Hölldobler and Wilson explain, ants are not especially difficult to observe and study.[57] Consequently, everyone with interest and patience can verify the humanlike description of ants just presented. We can predict a renaissance of research on ants as soon as the above data plus two additional facts become more widely known.

1. Scientists are told what they are supposed to think by the dogmas and commandments of official science. For instance, they are told implicitly or explicitly that ants are to be thought of as automata. Since true science is the search for verifiable knowledge without preconceptions and prejudgments, these official commandments, which intimidate almost all scientists, are totally unscientific.

2. Our greatest myrmecologists, such as William Morton Wheeler, have tried to tell us that ants shockingly resemble people in many significant ways—for instance, in their displays of pain, anger, fear, depression, elation, affection, and other emotions; in their apparent empathy when they help crippled and distressed nestmates; and in their ability to manipulate others by deception, such as by "playing dead."[58]

Honeybees

It is obvious that honeybees (*Apis mellifera*) have instinctual propensities for gathering pollen and nectar, building comb nests, moving close together in winter to keep warm by sharing their body heat, and carrying out all the other numerous activities that are intrinsic to honeybees. In essentially the same way as humans, honeybees implement their instinctual propensities by intelligent actions. Here are a few examples from the recent scientific literature.[59]

When the hive becomes crowded, some honeybees move out together in a coordinated swarm to find a new home.[60] First, they congregate in an organized way on a nearby tree. The bees most familiar with the environment, typically the eldest ones, serve as scouts and survey many miles of forest until they find one or more cavities that are appropriate for nest sites. The scouts behave in shockingly humanlike ways as they check out the cavities. They thoroughly inspect the inside of the cavity by walking or flying over every nook and cranny. They also examine the outside of the cavity by slow, hovering flights at increasing distances. They leave and return later to check the site during a different time of day. Elegant research has shown that the scouts carry out a process that humans mistakenly believe is unique to humans, that is, they evaluate or judge the desirability of the cavity as a potential nest on eight well-defined criteria: not drafty, dry, free from ants and other insects, facing south (to receive more sun in the winter), containing a hidden or easily defended nest entrance, large enough to store sufficient honey to survive the winter, small enough to be kept warm by the body heat of the bee colony, and far enough from the parent nest to allow enough food sources for both colonies.

After a long, meandering flight evaluating possible nest sites, the scout bee performs the outstanding feat of literally *navigating* (without instruments) directly back to the swarm by the shortest possible route. At the swarm site, the scout uses body language to inform the other scouts about the desirability of the cavities she has discovered and also about their location (their direction and distance). Highly esteemed researchers, beginning with Nobel laureate Karl von Frisch and continuing with Martin Lindauer, James Gould, and Thomas Seeley, have broken the honeybees' code and deciphered at least part of their body language.[61] They discovered that the distance, direction, and desirability of a new nest site is conveyed by a symbolic code that is embedded in the way the scout who has just returned to the swarm moves or "dances" in a patterned way. During these patterned movements, which have been termed a "waggle dance," the scout runs in a straight line while vibrating or waggling her abdomen a certain number of times, then returns to her starting point (via a clockwise semicircle to the right), runs over the same straight line while waggling again the same number of times, returns to her starting point (via a counterclockwise semicircle to the left), runs and waggles again, and so on.

The number of waggles per run symbolizes the distance to the potential nest site. However, honeybees in different locations assign different meanings to this symbolic code; the honeybee body language has different "dialects." Honeybees studied in Germany interpreted one waggle (per run) as indicating that the new nest site is at a distance of 45 meters; honeybees studied in Italy interpreted one waggle (per run) as meaning a distance of 20 meters; and honeybees studied in Egypt interpreted it as meaning 12 meters. Since honeybees represent or symbolize distance in space by the number of waggles per run, they resemble humans not only in their ability to represent or symbolize but also in their ability to count and to measure or judge distance!

The direction of the straight run part of the waggle dance in relation to the sun symbolizes the direction of the new nest site. The liveliness or vigor of the scout's dance movements represents the desirability of the new site.

After the returned scouts have reported the distance, direction, and desirability of their best nest sites to each other, they all fly out again to inspect sites reported by others. If another bee's find is more satisfactory than her own best find, the scout now "changes her mind" and, when she returns to the swarm, dances for the alternate site. All of the scouts now literally "vote with their feet" (by the vigor of their dancing), and the dancing-voting process continues until they all come to an agreement. Finally, the scouts guide the swarm directly to the agreed-upon new nest whose precise location had been originally communicated symbolically via the waggle dance language.

Although cognizant of the official science commandment that forbids any reference to humanlike mental processes in insects, a leading researcher in this area, Hubert Markl, courageously states that "it is hardly possible to describe what occurs in such a swarm without using vocabulary that implies a higher-order cognitive performance achieved during the process."[62]

The waggle dance language also can be used flexibly to report a new food source, the location of water, and the location of trees where a material used in construction (resin or propolis) is obtainable. Since these finds are reported to the bees in the interior of the dark nest where the sun is unobservable, the scouts have to change how they symbolize or represent the direction of the find. Inside the hive they flexibly change the code by using the angle of the straight run to the left or right of the vertical to indicate symbolically that the find is at the corresponding angle to the left or right of the sun.[63]

Although we have just begun to penetrate the honeybee's communication system, we already know that, like human language, it is symbolic in that it refers to things that are not present, it relies on agreed-upon rules for assigning meaning to the symbols, and its units can be rearranged in many ways to provide numerous different messages. A leading researcher, Thomas Seeley, writes that "a waggle dance is a truly symbolic message, one which is separated in space and time from both the action on which it is based and the behaviors it will guide."[64] Another foremost investigator, James Gould, agrees, writing that the honeybees' dance language is a "true language" that uses agreed-upon rules and refers to things that are distant in space and time. Gould adds insightfully that "Anyone who considers a communication system language only if it is wholly learned will dismiss the dance language, but . . . the same criterion, rigorously applied, also excludes human language."[65]

The dance language with its various dialects is the language of honeybees; other kinds of bees have different languages. For instance, several kinds of stingless bees (*Melipona* species) "use a kind of Morse code to indicate the distance to the food site"—a series of short sounds symbolize a nearby site while drawn-out sounds symbolize a greater distance.[66]

Returning to the honeybees, we find that the scouts do not proceed robotically to report their new food discoveries. On the contrary, they report them only when the colony needs additional food sources and when the quality and quantity of the new source is high and its distance not too great. Honeybees also are flexible in other aspects of their lives that have been investigated. "Construction workers" flexibly undo their earlier constructions when it is useful to do so; for instance, they may tear down part of a comb they have built in order to reconstruct it in a more useful way.[67] When combs are damaged experimentally, honeybees can rebuild them in ways "which none of their ancestors or none of themselves had ever built before."[68] Honeybees also flexibly change their work specialty to take on any of the jobs needed by the colony at a particular time; for instance, when more construction workers, or guards, or foragers are needed, these jobs will be promptly filled by workers who shift into them from other specialties. During hot weather, when water is needed to cool the hive by evaporation, many foragers shift to bringing in water and many workers within the hive take on the job of spreading a thin film of water over the larval cells or fanning their wings to create air currents that hasten evaporation.

Instead of robotically bringing in resin, pollen, and nectar and storing food for the winter, honeybees flexibly change their behavior to adapt to sudden changes in the environment. They stop collecting gluey resin from trees when a man-made substitute is nearby. When pollen is scarce, they will substitute oatmeal or flour that beekeepers place near the hive. When a sugar refinery was built near a hive, honeybees used the sugar and stopped foraging for nectar. Within two years after northern bees were transported to a tropical location, where flowers are always plentiful, they stopped storing provisions for a winter that would not come.[69]

James Gould recently discovered a simple test of intelligent awareness applicable to bees. Following an earlier suggestion by Karl von Frisch, he asked: If food for honeybees (sugar solution in a dish) is systematically moved at regularly increasing distances away from the hive, do the bees "catch on" to the regular movements and wait for the sugar at the next scheduled location? Gould reports that he and his students observed an "eerie phenomenon" in which "some of the bees began to anticipate our movements . . . fly on past the training station and wait at the spot we are aiming for on the next move."[70] These results appeared eerie because the bees behaved essentially like intelligent humans; they insightfully "caught on" to the experimenter's logical, systematic movements and could predict them and use them for their own benefit.

The tiny honeybee is superior to you and me in natural navigational intelligence. Its skill in navigating equals that of the most proficient Polynesian navigator who has undergone extensive training and supervised practice. The accumulated research in this area that was initiated by Nobel laureate Karl von Frisch converges on the conclusion that the very young bee, like the very young bird and the very young ant, learn quickly and easily to "read" compass directions from the sun by observing and remembering the sun's apparent motions across the sky and relating these motions to the steady passage of time and to geographic directions. Like birds and ants, honeybees also can use other natural compasses, especially when the sun is not visible; these include the patterns of polarized light in the sky and magnetic patterns on the earth's surface. Using one or more of these natural compasses and a sense of the passage of time, the honeybee is able to keep track of the directional angle and distance of each leg of its outward journey and to integrate this information to determine the shortest direct line back to its hive.[71] The navigational intelligence of honeybees (and birds and ants) appears to be a special intelligence in essentially the same way as the special linguistic-symbolic intelligence of humans and the special musical intelligence of birds, humans, and whales.

The animals discussed in this book—humans, birds, gorillas, chimpanzees, dolphins, whales, fish, ants, and bees—are among the ones that have been most thoroughly investigated. The hard scientific data indicate that they are intelligently aware. Since the animals considered can be viewed as representative of both vertebrates and invertebrates, it appears likely that *all* animals are intelligently aware. It is now incumbent upon investigators to test this deduction for each of the earth's taxa.

CHAPTER TWELVE

Revolutionary Implications of Animal Intelligence

As people begin to see willful intelligent awareness where they previously saw only machinelike processes, their relationship to nature and the universe will change drastically. As the humanlike qualities of birds and other animals penetrate deep into the consciousness of a new generation, humanity's philosophy of life will turn around along with human cultural institutions. Science, religion, and philosophy will be fundamentally different. No longer will scientists assume that humans are the only intelligent life on planet earth; on the contrary, the next generation of scientists will be increasingly aware of the conscious, intelligent life that covers the surface of the earth, beginning with the nearby birds and extending out to the other living things. No longer will religion focus only on God and humans while disregarding the other life on the earth.

Instead, as the theologian Philip Sherrard recently predicted, the new religion will regard nature and its creatures "not as something upon which God acts from without," but as "something through which God expresses Himself within [and is] totally present within."[1]

No longer will philosophers dare to philosophize without ever mentioning any of the earth's nonhuman animals. Instead of being separated from and elevated above the rest of nature, *Homo sapiens* will be seen to be one of the earth's countless creatures, as sensitively aware, as intelligently competent, and as specialized as many others. (The specialized ability of humans to use symbols and tools will be seen to have recently produced their vaunted technology, their conquest of the earth, and their capability to degrade the earth's atmosphere, hydrosphere, and biosphere.) As people realize the humanlike qualities of birds and other animals, a new respect and reverence will enter into their relationship with them. Their philosophy of life will more closely resemble that of "primitive" people, such as the Hopi and the Eskimo, who approach all life (including the plants or animals they take for their sustenance) with respect and reverence and with a basic understanding of the essential ecological principle that underlies the continuity of life on earth—the principle of perpetual recycling of life from plants, to herbivores, to carnivores, to decomposers and back to plants with everything, in the final analysis, interdependent and feeding directly or indirectly on everything else.[2]

The idea of avian intelligence, which has been the focus of this book, will be resisted, as are all revolutionary conceptions, by a large segment of scientists and nonscientists who unquestioningly accept "what everyone knows is true"—the assumptions and propositions of the dominant paradigm.[3] However, the essential commonality of birds and humans can be demonstrated unequivocally in various ways. Following in the pioneering footsteps of Alex the parrot and Blue Bird the parakeet, befriended birds in the future can demonstrate their humanlike awareness and understanding by conversing meaningfully and intelligently with humans (as if they too are humans) about the immediate things that matter — for instance, "I want some tea," "Come here," "You tickle me," "Go pick up cup," "Go away," "Open the door," "Wanna go gym," "Wanna go back," "Take a shower," "Can I have some?" "You tell me, What color?" "How are you?" "What are you doing?" "Where are you going?" "Hello," "Good-bye," "Good morning," "Good night," "Give me a kiss," "Pretty little Blondie . . . Blondie's so nice." The human-like qualities of wild birds also can be demonstrated unequivo-cally by thorough visual documentation of individual birds. Such documentation will negate the pseudoscientific dogma that birds are automata and show them as they are — unique and complex personalities sensibly and proficiently singing, mating, nest building, parenting, navigating, playing, emoting, socializing, and constantly experiencing.

Since avian intelligence and awareness is a factual, demon-strable conception, it cannot be squashed and is bound to pre-vail. How quickly humanity's conception of reality changes will depend to a significant degree on how profoundly and personally readers of this book understand the sensitivity, awareness, and intelligence of birds and how effectively the readers communi-cate this understanding to others. The forthcoming revolution in human thought will be led by men and women who are no longer intimidated by the taboo against perceiving birds as conscious individuals; who devote much time and effort befriending birds in the wild and raising birds freely in their open homes (while employing safeguards such as those discussed in Appendix B); who guide birds to bond with them and learn their language; and who transmit to others (via lectures, demonstrations, television shows, news articles, journal articles, and group projects) their perception of the humanlike characteristics of birds. The avian revolution will be complete when the new generation accepts as natural that people and birds can understand each other and relate to each other not only as equals but also as friends. The revolution will have triumphed when reports such as the follow-ing, by Len Howard, are neither strange nor unusual:

A great many people have shown interest in my tame-wild birds. Often strangers see me with them perched on my hand down the road and stop to ask how it is these birds have got so tame. In this "better world" now being planned for future generations will it still be so rare to see the beautiful wild birds perching fearlessly on human hands? I always remember the words of an electrician who once called to attend to fittings in my Sussex cottage. He stopped in amazement before my doorway, watching countless birds flying down from the trees to perch on me. He had looked an ordinary man with a work-a-day expression until he saw these birds, then his whole countenance seemed to alter, his face glowed, his eyes shone and he kept murmuring: "How wonderful!" Then he said: "But why shouldn't it be like that? It ought to be like that."[4]

Humanity, the Destroyer
◆

The call for a revolution in humanity's conception of reality is urgent. At the present time humans are harming everything in nature, including most obviously their closest wild neighbors, the birds. Many kinds of birds have already been totally annihilated; numerous other species are in danger of extinction, including the golden eagle, the southern bald eagle, the California condor, the black-necked swan, the whooping crane, and so on and on. The birds around our homes are fewer and fewer each year. In areas of the United States where there were many birds thirty or forty years ago, there are hardly any today. The threat to the avian population is moving so fast that serious scholars are beginning to doubt whether the next generation of Americans will know what it means to hear a wild bird sing.[5]

Humans think nothing of their destruction of birds. Very, very few humans are much concerned that they are poisoning birds by poisoning the air they breathe, the water they drink, and the food they eat with herbicides, insecticides, industrial gases—

carbon dioxide, carbon monoxide, sulfur dioxide, nitrogen oxides, and so on and on. How many people really care that their species has already destroyed the habitats and thus extinguished many species of birds and is now destroying the tropical forests where nearly half of the nine thousand or so different kinds of birds make their living? The cries of anguish of decent human beings over the genocidal destruction of the Tasmanians and the near destruction of the American Indians in their native lands, of the Armenians and Greeks in Turkey, and of the Jews in Germany will soon be followed by cries of anguish over the genocide of the many different species and genera of birds that is occurring worldwide.

There are three major reasons why each of us should do everything we can to inform every person who will listen of the humanlike qualities of birds.

First, it is our duty to save bird species from destruction. This obligation is embedded in a higher ethics that transcends and supersedes humanity's traditional anthropocentric, self-centered ethics—a loftier reverential ethics that values each form of life as sacred and sees genocide (of human groups or nonhuman species) as an unforgivable sin. This reverential ethics requires treating all forms of life with respect. Such respect for a natural order comprised of omnivores, carnivores, and herbivores can be seen, for example, when ancient human hunter-gatherers consumed the flesh of flora and fauna only as needed for their basic nutrition and with reverence and appreciation for the sacrificial plant or animal.

Second, humans are not only poisoning the birds; they are simultaneously poisoning all of the earth's flora and fauna at an accelerating rate. By working to save the birds we humans are ipso facto also working to save the other animals and plants and the earth itself from mankind's degradation.

Third, by ruining the health and life of birds and other animals and plants, humans are also ruining their own health and quality of life. If humans save the birds, they also will save the other life on earth, including their own. The pollutants, toxins, and poisons that are destroying birds affect every form of life, including human beings. For humans to save themselves, however, they need a fundamental change in consciousness, a new understanding, a realization that they are not the only intelligent beings on earth—that birds, as a prime example, are as aware and sensitive and have as much practical intelligence as they. Humans have to see the human race not as the appointed ruler of the earth but as one of its innumerable specialized species. Humans have a responsibility to use their intellectual-technological speciality wisely without harming other species and for the benefit of all, in essentially the same way as individual people specializing in intellectual and technological pursuits (such as scientists and engineers) have a responsibility to use their specialties without harming other people and for the benefit of all.

Very few people are as yet aware of the most crucial fact of all: Humanity's waste products—feces and garbage, and agricultural, vehicular, military, and especially industrial wastes—are accumulating in the water, air, soil, forests, and food chain faster than they are being removed by natural processes.[6] This excess accumulation of unnatural products is unhealthy for human beings and all other organisms. Health declines as environmental contaminants increase. Life on earth cannot remain healthy when the air, water, and soil are increasingly burdened with toxins, poisons, pesticides, herbicides, insecticides, endless lists of industrial pollutants, and dangerous vehicular exhausts.

Humans are polluting the earth's groundwater, brooks, creeks, rivers, lakes, and seas at an alarming rate. All of the groundwater aquifers in Long Island have been poisoned, one-third of those in Massachusetts, and one-fourth of those in California. The Rhine River ecology has been nearly destroyed by eight tons of pure mercury and a thousand tons of other dangerous chemicals that have entered the river (much from a single huge fire in a chemical depot). The Mediterranean Sea is being polluted at such an accelerating pace that it now supports fewer than half the fish it contained just twenty years ago. Water contamination is increasing with advancing high-tech industries;

production of the critical computer component (the silicon chip), for instance, is contaminating the water with toxic cyanides, chemical solvents, arsenic, and heavy metals.[7]

The balance of gases in the atmosphere which is necessary for life is being changed dangerously by an increasing number of man-made contaminants. Toxic chemicals emitted by American industries in the form of solid particles, invisible gases, and aerosol mists are now accumulating in the atmosphere at the rate of 1,400 pounds per year for each American.[8] These industrial air contaminants are supplemented by a huge mass of pollutants from automobiles, agriculture, and the military.[9] The accelerating production of atmospheric pollutants has already significantly changed the air that is breathed by birds, people, and other animals, especially in industrial and urban areas. It has already affected the health and even the lives of the most vulnerable, such as young songbirds and elderly people.[10] Chlorofluorocarbons and other man-made chemicals are reducing the effectiveness of the atmospheric ozone layer in protecting humans and other animals from the negative consequences of excessive ultraviolet light (skin cancers, cataracts, and damage to the immune system). The rapid atmospheric increase in heat-absorbing gases, such as carbon dioxide, methane, and nitrous oxide, especially from industry and deforestation, is producing a significant heating of the earth's surface; in order of increasing temperature, the eight warmest years in the past 121 years (from 1870 through 1991) were 1980, 1989, 1981, 1983, 1987, 1988, 1991, and 1990. Serious consequences from the increasing heat are expected, including instability in the global climate with more frequent and powerful tornadoes and hurricanes and accelerated melting of the polar icecaps with subsequent rising sea levels and flooding of coastal areas.[11]

Every few years the atmosphere is contaminated by man-made lethal radiation; witness accidents at nuclear plants at Three Mile Island in Pennsylvania in 1979 and Chernobyl in the former Soviet Union in 1986. Many thousands of people within six hundred miles of Chernobyl received cancer-producing doses of radiation from the original release of radioactive material. Many more, including people in the United States, were unknowingly affected as the radioactive material moved in a cloud over the earth and was brought down by rain.[12] Of course, uncounted numbers of innocent birds and other animals were also killed or maimed by the Chernobyl release of radioactivity[13] and by other recent man-made disasters, such as the release of poisonous chemicals from the American-owned plant in Bhopal, India, that killed between 2,000 and 3,000 people and injured 200,000.[14]

By 1988 the U.S. Environmental Protection Agency had succeeded in "cleaning up" only twenty-two of the fifty thousand dump sites that were leaking poisons into the soil and groundwater of the United States. When the effects of such leaking dumps were carefully investigated at one site (Love Canal in New York), it was discovered that of eighteen pregnancies occurring there, two resulted in normal children, nine in serious birth defects, and seven in stillbirths and spontaneous abortions. Alarmingly, birth defects among humans have doubled during the past twenty-five years and have increased exponentially among birds and other animals.

Humanity's accelerating degradation of the earth's life support systems is reaching unbelievable levels. The rich soil of Iowa is being depleted by devastating agricultural practices at a rate that, if continued, would render it a desert in about thirty years! The best estimate now is that the earth's fertile soil will be reduced by one-third in fifteen years.[15] The tropical forests, which are crucial for the life of more than half of the earth's species and which are surprisingly important in regulating worldwide temperature and rainfall, are being destroyed at a rate that will render them deserts in a few decades. Tragically, the nutrients in these tropical forests are so close to the surface that the cleared land can support crops only for a few years. Then it turns into a desert.

Humanity's direct environmental devastation is just one aspect of the multifaceted devastation being wreaked on the animals, plants, atmosphere, and hydrosphere. Increasingly suicidal wars regularly destroy not only entire human populations but also vast numbers of nonhuman beings. The fast-increasing number of humans on the planet, especially in the poverty-stricken nations of South America, Africa, and Asia, is outdistancing the availability of local resources. The best estimates are that more than one billion people, about one out of every five people on the earth, are suffering daily hunger and that in time unimaginable numbers of them will die directly or indirectly from the effects of malnutrition. The driving need of our present industrial system to constantly increase production ("progress") virtually guarantees such negative consequences as resource depletion, species extinction, wilderness destruction, deforestation, and desertification.

Humanity Renewed
◆

Humanity's destructive impetus will be stopped when the natural world is perceived in a new way. After an intense period of deep thinking on these topics, the scholar Thomas Berry arrived at three very important conclusions.

◆ To commit to the effort needed to stop the destruction, people need a comprehensive vision—a radical change in their sense of reality.

◆ For this radical change to occur, people have to see themselves as they truly are—one species among many in a larger community of life. They have to experience a "reenchantment" with their kin in the earth community, reestablish themselves within a natural context, and integrate their well-being with the well-being of the natural world. Only within the ever-renewing recycling processes of nature is there a healthy future for the human community.

◆ The most important task now is to guide the new generation to see the grandeur, meaning, and sacredness of the natural world.[16]

After careful reviews of the research on environmental devastation, other thinkers recently arrived at very similar conclusions.

♦ "For humanity to be part of the peace process, to
   cease being warriors against Mother Earth, a great
   awakening must take place."—Matthew Fox[17]

♦ To prevent further ecological destruction and the
   possible end of our civilization, "a quasi-religious
   movement, one concerned with the need to change
   the values that now govern much of human activity, is
   essential."—Paul R. Ehrlich[18]

♦ "To survive we face the hard task of . . . learning
   again to be part of the Earth and not separate from
   it."—James Lovelock[19]

♦ To solve the horrendous environmental problems that
   are upon us, "all our old habits and vested interests,
   even if they form our individual and national identity,
   must be fundamentally changed. The changes required are deeper and more far-reaching than any
   revolutionary leader has ever demanded or even
   dreamed of demanding."—Elisabet Sahtouris[20]

The deepest and most far-reaching change that is demanded is
for human beings to realize that they are but one kind of life form
on the earth, as useful and specialized as all the others; that their
intellectual-technical specialty does not give them the right or
privilege to dominate and enslave other species; and that they
have no more inherent kingship over the other life on the planet
than specialized scientists, intellectuals, and engineers have over
athletes, firefighters, and other humans with nonintellectual,
nontechnical specialties.

As long as people see themselves as having been given
dominion, the inherent right to control the earth and its species
and to do with them whatsoever they wish, they will continue to
pollute and degrade the earth and to bring sickness and death to
the earth's species. As soon as people see that they are as good

and as useful as every other earth creature and that their symbol-tool specialty is as awesomely impressive as the specialties of other species, they will turn away from the road of destruction.

People of the earth, awaken! Open your eyes, look around you, and become aware of the fast-moving lives of your neighbors, the birds. Like you they are enjoying, playing, working, parenting, building, singing, socializing, loving, hurting, feeling, worrying, communicating, planning. Look closer and see the strivings, feelings, experiences of the individual animals near you. Wake up! Realize that you are as wonderfully aware, as fully conscious, and as specialized as the other creatures on the earth. Use your specialized intelligence now to change your destructive habits, to save the earth's flora and fauna, including yourself, from further devastation, and to live in harmony with deep enjoyment.

The Continuing Cognitive Ethology Revolution

◆

A CLARIFICATION FOR MY COLLEAGUES IN THE BEHAVIORAL AND BRAIN SCIENCES

This book is an offshoot of the cognitive ethology movement, which postulates mental experiences in nonhuman animals and aims to transform comparative psychology, ethology, and related behavioral and brain sciences. Cognitive ethology was founded and is led by the highly esteemed scientist Donald R. Griffin, professor emeritus at Rockefeller University. In many papers and two ground-breaking books— *The Question of Animal Awareness* (1976) and *Animal Thinking* (1984)—Griffin synthesized for the scientific community the results of several dozen carefully conducted research projects carried out primarily with mammals, birds, and insects that indicate flexibility, awareness, simple feelings, and simple thoughts in nonhuman animals.[1] From the accumulated data Griffin deduced a series of important conclusions including the following:

Animals appear to guide their behavior by simple cause-effect thinking: "If this, then that"; "If I dig here, I will find food"; "If I dive into my burrow, that creature won't hurt me"; "If I [a lab pigeon] peck at that bright spot, I can get grain"; "If I [a lab rat] press the lever, the floor won't hurt me."[2]

If animals actually think, ethologists, comparative psychologists, and other students of animal behavior must change their fundamental assumptions. Since the question of animal mentality is of basic scientific importance, scientists should address it seriously, without prejudice, and with all their energy and resources.

Unfortunately, scientific inquiry into this important area has been blocked for many years and continues to be seriously hindered today by the behavioristic-reductionistic-positivistic *Zeitgeist*. Under this dominant paradigm, students learn that it is unscientific to ask what an animal feels or thinks; research scientists fear ridicule and ostracism if they interpret data as showing animal awareness; field naturalists hesitate to write publicly about the mentality of the animals they study; and editors are quick to reject papers that interpret data as indicating animal awareness or cognition.

◆ The most conservative reasonable conclusion that can be drawn from the accumulated data is that scientists should strive to free themselves from the strictures of the dominant paradigm and look again with an open mind at the possibility of animal mentality.

Griffin's work was the stimulus for the present book. After thirty years and 180 publications as an active research psychologist, I decided to test Griffin's program by carefully observing my closest wild neighbors, the birds, and by reading every significant scientific book and journal article that has been written about avian behavior during the past thirty years. The voluminous data I encountered during my six years of total immersion in avian studies and observations led to the unexpected conclusion that not only are birds able to think simple thoughts and have simple feelings, but they also are fundamentally as aware, intelligent, mindful, emotional, and individualistic as ordinary people. This conclusion not only went beyond the thinking birds portrayed by Griffin, it also strikingly violated the basic commandment of the dominant paradigm — Thou Shalt not Anthropomorphize!

Since this book contradicts the anti-anthropomorphic basis of the dominant paradigm, it is predictable, from the writings of Thomas Kuhn, Imre Lakatos, Larry Laudan, and others who have analyzed the conduct of challenged scientists,[3] that committed adherents of the dominant view will strive to remove it from scientific consideration by ridicule and direct and indirect attacks.[4] However, it is also predictable that in the long run the revolutionary anthropomorphic view of birds will become dominant because, as indicated by the data in this book, it can be empirically demonstrated that birds have the characteristics that people thought were unique to people and as these demonstrations become increasingly more comprehensive and powerful, a new generation of scientists and laypeople will accept the essential human-avian similarity as a basic scientific truth.

Note added in proof. During the time *The Human Nature of Birds* was in press, Griffin published a new book — *Animal Minds* (Chicago: University of Chicago Press, 1992) — which amplifies and updates his earlier books.

◆

APPENDIX B

How You Can
Personally
Experience a Bird
as an Intelligent
Individual

You too can befriend birds, look into their secret lives, and perceive their hidden personalities. Depending on your life circumstances, you and your children and friends may be able to follow the procedures pioneered by Len Howard with wild birds or by Sheryl C. Wilson with birds living as free individuals in your home.

Befriending Wild Birds

◆

As described in Chapters 6 and 8, Howard was an English musicologist who moved to a country cottage to study the song of birds and was surprised to discover that she was able to form close friendships with numerous wild birds and get to know them as well as she knew people. She observed the secrets of their lives and learned that basic avian nature and basic human nature include essentially the same components—pain, happiness, sadness, joy, playfulness, sexual enjoyment, mindfulness, awareness, and practical intelligence. She recognized the profound truth that birds are essentially like people because she approached them with an open, calm, respectful attitude. Her life soon consisted of continual interactions with birds. Here is how she described it: When an ordinary workingman saw

> the beautiful wild birds perching fearlessly on [my] hands . . . countless birds flying down from the trees to perch on me . . . his face glowed, his eyes shone and he kept murmuring: "How wonderful!" Then he said: "But why shouldn't it be like that? It ought to be like that." . . . living as I do, in continual company of numbers of birds . . . they do all they can to prevent my concentrating upon anything except themselves. . . . But their lives are short and there are many tragedies. . . . Most mornings recently I have been awakened at five o'clock by a Great Tit making agitated flights to and from my bed to the window while uttering loud alarm cries. He is telling me to come out quickly, the Magpie is endangering his young, so I leap from bed and chase off this enemy with a stick. I return to bed, but soon there is more trouble, the Blackbird calls me up by agitated "tchinks" close to my window, and again I go out to frighten away the cat by flinging a jug of water at it. . . . If I go on a holiday so many disasters occur that I very rarely go away, although wanting to watch other kinds of birds further afield. . . . In one way and another my birds demand attention from dawn to dusk.[1]

How did Howard become so intimate with the birds around her? She tells us:

Perhaps it is because of my intense love for birds that they come to me quickly and I have not found any difficulty in gaining their confidence. . . . I put up a bird-table and bath close to the french window, and a Robin, Blue Tit, and Blackbird came at once, many more species . . . soon following. . . . Very quickly this great intimacy developed [especially after offering special treats to the birds by hand] and the numbers rapidly increased. Besides loving their company, I find immense interest in studying their individual characters. . . .[2]

When Howard first approached the birds, she "was not expecting much intelligence to be shown in their behaviour."[3] Consequently, she was surprised to discover that they consistently acted intelligently. She writes that "Often bird behaviour is judged when the bird is panicked with fear of the watcher. . . . I find the normal thing is for birds . . . to act intelligently in unusual circumstances unless they get flustered through fear."[4]

Howard was also pleasantly surprised to find that knowing birds and interacting with them was more fun than interacting with most people. Besides the joy she received from her avian friends, Howard also gained deep aesthetic satisfactions, for instance, by observing the flight of birds, which she wrote about as follows:

Birds not only use flight as a natural means of locomotion, but in beautiful forms as a means of expression. . . . Many species spend hours of the day in the recreation of flight, as others spend hours in song. Flight is an art akin to music, with rhythm and feeling of movement as its foundation, a glorious means of expression that birds, with their emotional natures, know well how to use. Some species . . . have developed flock-flight in unison to such perfection that the inquisitive human is forever wondering how they achieve this simultaneous movement without a conductor to assist. [Using her own personal experiences as a musician, Howard realized that the experience of birds during harmonious flock flight maneuvers must be akin to the experience of musicians who play together harmoniously without a conductor because they are] "keyed up to a quickened response . . . [and] feel together as one. . . . All are swayed by the same impulse or inspiration, and each feels a supersensitive consciousness of the other player's interpretation, often to the point of feeling it slightly in advance . . . and they are not only thrilled themselves, but their audience can feel the thrill of their performances."[5]

Howard also described her aesthetic feelings for the singing of birds: "The more one listens to their music, gathering knowledge and understanding of their musical language, the more beauty the subject holds. Song is an emotional outlet, a bird's heart goes into his music. . . ."[6] "Musical talent varies individually—within species—as much as among human performers of music. . . . This variation of talent is not only a question of voice quality, but such things as the material of the song or musical composition; the interpretation of the composition, and the technical ability. . . ."[7]

The Willow-Warbler's song . . . goes straight from heart to heart, where there is true sympathy between bird and man. Each tone seems to be delivered with loving care, especially after nesting begins. The rapidity of utterance and strength of tone sound thoughtfully graduated as the pure, sweet notes gently fall, then rising a little, fall again to a softer cadence. . . . [The Nightingale's] mysterious note, repeated on a crescendo, is both wonderful for its emotional effect and for masterly control of technique. . . . Forceful rhythmic phrases and bubbling trill crescendos are performed with amazing skill and driving power . . . then comes . . . a pause that is part of the song, when the stillness and beauty of night are felt with growing intensity, then the wonderful note is heard and it seems sprung from the poetry of night.[8]

By implementing Howard's three major procedures (see p. 88, and following), you too can experience the joy and aesthetic satisfaction she derived from understanding the hidden life of birds.

Place bird food around your home wherever birds can see it and eat it—on the floor of your balcony, on your window ledge or deck or patio, on the bare ground or on trays or bird tables.[9] Also, place bird baths and nest boxes around your abode. (Later, whenever possible, place food and nest boxes in parts of your home open to the outside.)

Approach the birds with the calm, open, happy, positive frame of mind that is you at your human best.

Begin to offer food that the birds can take from your hand so that they can approach closer to you and bond with you.

Befriending Birds Living Freely in Your Home

◆

To obtain a close, personal view of the intelligence and personality of a bird living in your home, use the methods pioneered by Sheryl C. Wilson (see Chapter 8). In finding and raising the parakeet Blue Bird, she proceeded as follows[10]:

After reading about the characteristics of various kinds of birds, Wilson chose a budgerigar (parakeet) because they are reputed to be social and playful, they have the potential for learning to talk, their food is readily available, their stainless and odorless droppings are easy to remove, and they are inexpensive (about twenty dollars).

Wilson purchased a home-raised parakeet from a local bird breeder at the ideal age—five to six weeks after hatching, when a parakeet is ready to leave its parents and to form a relationship with a human. From a number of parakeets at five to six weeks of age, she picked Blue Bird because he was the most active and seemed curious and interested in her.

Wilson was aware that it could be physically and emotionally stressful for Blue Bird to be taken away from his home, his parents, and the other parakeets. To minimize trauma when transporting him by taxi to his new home, Wilson held his small box (containing breathing holes) securely on her lap and spoke to him soothingly to reassure him during the short journey.[11]

Upon arrival in his new home, Wilson reassured Blue Bird that he was safe in the same way she might have reassured a frightened child from another culture, for example, by approaching him with a friendly and calm attitude and addressing him by his name spoken in a soft tone of voice. During Blue Bird's first week or so in his new home, Wilson provided peaceful surroundings by removing stimuli that might frighten him (such as the loud and strange sounds of a vacuum cleaner),[12] and she spent many hours each day in his presence so that he could observe her carrying out routine, nonthreatening activities, such as eating, reading, writing, and softly playing the piano. Approaching him slowly and gently so as not to startle him, she also offered special food treats, such as spray of millet, by hand so that he felt increasingly at ease with her.

Soon Blue Bird was secure in his new home, and it was time to let him out of his cage to fly freely. Doors, windows, and other exits were closed so that he would not fly out of the house.[13] Mirrors and windows were covered so that he would not fly into them. (Later he was carefully introduced to their reflective properties.) Also, common sense was used to remove possible hazards. Methodically looking around each room he would fly in, Wilson asked: What could burn him? (Hot pans and pots, the stove, the electric heater, the metal lamp.) Where could he drown? (In the standing dishwater, the open toilet bowl, the partially filled water glass.) Where could he get stuck? (Cupboards, drawers, vases, the spaces between the furniture and the wall.) How could he be safe from pollutants? (By removing aerosols, artificial air fresheners, strong cleansers, insecticides, pesticides, and using only ecologically safe products.) What else might harm him? (Poisonous plants, sprayed vegetables, lead from pencils, electric wires.)

Since Wilson had read widely about parakeets and had

discussed specifics with the breeder, she knew how to meet Blue Bird's needs for socialization (by playing, talking, and interacting frequently with him) and how to fulfill his requirements for certain foods, toys to play with, and places to play in and things to perch on when away from his cage (a playground or gym and a "parakeet tree").[14]

Blue Bird learned to talk meaningfully during social interactions with Wilson. In the appropriate contexts, she spoke phrases that were useful to him, such as "Open the door," which he could later use when he wanted his cage door opened. (Professor Pepperberg's social modeling procedure can also be used to teach a parakeet or parrot to talk meaningfully. As noted in Chapter 1, two people serving as models say things to each other that are of interest to the observing bird; for example, "What's this?" [ball] and "What color?" [orange] are phrases a bird can use later to request an orange ball.)[15]

The Importance of Befriending Birds

◆

By working with devotion and commitment to befriend birds and to understand their intelligence and personalities, you will be acting as part of a revolutionary movement that will change the consciousness and destiny of the human race. As you and your compeers succeed in personally recognizing the fact of avian intelligence, you will naturally organize to spread this fact to others. With the potent communication media now available, it will be possible to demonstrate immediately and convincingly to people worldwide that your bird has the essential characteristics that people have (wrongly) believed were unique to people. As soon as humankind sees that it has totally misjudged the basic nature of its closest wild neighbors, the birds, it will be shocked out of its lethargy (which is leading to a devastated and hopeless world), it will start thinking anew, and it will move to a truer and deeper understanding of reality and to a better way of life.

Scientific Names of Species

Albatross, Laysan	*Diomedea immutabilis*
Blackbird, European	*Turdus merula*
Bowerbird, orange-crested gardener	*Amblyornis subalaris*
Bowerbird, satin	*Ptilonorhynchos violaceus*
Bunting, indigo	*Passerina cyanea*
Buzzard, black-breasted	*Haemrostris melanosterna*
Cacique	*Cassicus cella*
Canary	*Serinus canaria*
Chaffinch, common	*Fringilla coelebs*
Chicken	*Gallus gallus*
Cock-of-the-rock	*Rupicula rupicola*
Crow, common	*Corvus brachyrhynchos*
Crow, hooded	*Corvus cornix*
Cuckoo, common	*Cuculus canorus*
Duck, mandarin	*Aix galericulata*
Falcon, peregrine	*Falco peregrinus*
Finch, woodpecker	*Camarhynchus pallidus*
Goldfinch, European	*Carduelis carduelis*

Goose, Canada	*Branta canadensis*
Grouse, black	*Tetrao tetrix*
Grouse, sage	*Centrocercus urophasianus*
Gull, laughing	*Larus atricilla*
Gull, ring-billed	*Larus delawarensis*
Hawk, broad-winged	*Buteo platypterus*
Heron, green-backed	*Ardeola striata*
Jackdaw	*Corvus monedula*
Jaeger, pomarine	*Stercorarius pomarinus*
Jay, blue	*Cyanocitta cristata*
Jay, green	*Cyanocorax yncas*
Jay, pinyon	*Gymnorhinus cyanocephalus*
Jay, scrub	*Aphelocoma coerulescens*
Kingbird, Eastern	*Tyrannus tyrannus*
Kingfisher, ruddy	*Halcyon coromandus*
Kite, black	*Milvus migrams*
Linnet	*Carduelis cannabina*
Mallard	*Anas platyrhynchos*
Martin, house	*Delichon urbica*
Martin, purple	*Progne subis*
Mockingbird	*Mimus polyglottos*
Nightingale	*Luscinia megarhynchos*
Nutcracker, Clark's	*Nucifraga columbiana*
Nuthatch, North American	*Sitta pussila*
Nuthatch, pygmy	*Sitta pygamae*
Owl, great horned	*Bubo virginianus*
Owl, tawny	*Strix aluco*
Parakeet (Budgerigar)	*Melopsittacus undulatus*
Parrot, African grey	*Psittacus erithacus*
Parrot, kea	*Nestor notabilis*
Peacock (male peafowl)	*Pavo cristatus*

Pheasant, Argus	*Argusianus argus*
Pigeon, domestic ("laboratory")	*Columba livia*
Pigeon, tooth-billed	*Didunculus strigirostris*
Raven	*Corvus corax*
Robin, European	*Erithacus rubecula*
Rook	*Corvus frugilegus*
Ruff	*Philomachus pugnax*
Shearwater, greater	*Procellaria gravis*
Shearwater, manx	*Procellaria puffinus*
Shrike, bou-bou	*Laniarius aethiopicus*
Shrike, red-backed	*Lanius collurio*
Sparrow, song	*Melospiza melodia*
Starling	*Sturnus vulgaris*
Sunbird, golden-winged	*Nectarina reichenowi*
Swallow, barn	*Hirundo rustica*
Tailorbird	*Orthotomus sutorius*
Tern, Arctic	*Sterna macrura*
Tit, blue	*Parus caeruleus*
Tit, great	*Parus major*
Titmouse, penduline	*Remiz pendulinus*
Turkey	*Meleagris gallopavo*
Vulture, Egyptian	*Neophron percnopterus*
Warbler, black-throated	*Gerygone palpebrosa*
Warbler, Cape May	*Dendroica tigrina*
Weaver, village (black-headed)	*Ploceus cucullatus*
Woodpecker, tropical	*Micropternus brachyurus*

◆

NOTES

The numbers within brackets following an abbreviated title refer to the note in the same chapter in which the complete citation was originally presented.

Chapter 1. Avian Intelligence

1. I will at times refer to birds in general and to humans in general even though such generalizations always need qualification. For instance, only some birds and some humans realize the musical potential of *Aves* or *Homo sapiens.*

2. The meaning I assign to the term *intelligence* has three facets that are often forgotten:

   a. As emphasized by Donald R. Griffin, the founder of cognitive ethology, intelligence is present when goals are met by flexibly varying behavior while implicitly taking account of cause-effect relations ("If this, then that," "If I do a particular thing, another particular thing will happen.") (See Appendix A.)

   b. There are various kinds of specialized intelligences, including linguistic intelligence (a specialty of humans), musical intelligence, spatial intelligence, and bodily kinesthetic intelligence (which may also be highly developed in other animals). See Howard Gardner, *Frames of Mind: The Theory of Multiple Intelligences* (New York: Basic Books, 1983).

   c. As Freud and many others have correctly emphasized, only a tiny part of the intricately complex mental processes of humans (and by extension of other animals) is ever conscious or in the foreground of awareness.

3. When the text refers to a particular avian species by its common name, its scientific name is listed in Appendix C.

4. I. M. Pepperberg, "Functional Vocalizations by an African Grey Parrot (*Psittacus erithacus*)," *Zeitschrift für Tierpsychologie*, 1981, *55,*

139–160; I. M. Pepperberg, "Cognition in the African Grey Parrot: Preliminary Evidence for Auditory/Vocal Comprehension of the Class Concept." *Animal Learning and Behavior*, 1983, *11*, 179–185; I. M. Pepperberg and F. A. Kozak, "Object Permanence in the African Grey Parrot (*Psittacus erithacus*)," *Animal Learning and Behavior*, 1986, *14*, 322–330; I. M. Pepperberg, "Acquisition of Anomalous Communicatory Systems: Implications for Studies on Interspecies Communication," in R. J. Schusterman, J. A. Thomas, and F. G. Wood (Eds.), *Dolphin Cognition and Behavior: A Comparative Approach* (Hillsdale, N.J.: Lawrence Erlbaum Associates, 1986), pp. 289–302; I. M. Pepperberg, "Evidence for Conceptual Quantitative Abilities in the African Grey Parrot: Labeling of Cardinal Sets," *Ethology*, 1987, *75*, 37–61; I. M. Pepperberg, "Acquisition of the Same/Different Concept by an African Grey Parrot (*Psittacus erithacus*): Learning with Respect to Categories of Color, Shape, and Material," *Animal Learning and Behavior*, 1987, *15*, 423–432; I. M. Pepperberg, "An Interactive Modeling Technique for Acquisition of Communication Skills; Separation of 'Labeling' and 'Requesting' in a Psittacine Subject," *Applied Psycholinguistics*, 1988, *9*, 59–76; I. M. Pepperberg, "Evidence for Comprehension of 'Absence' by an African Grey Parrot: Learning with Respect to Questions of Same/Different," *Journal of the Experimental Analysis of Behavior*, 1988, *50*, 553–564; I. M. Pepperberg, "Cognition in the African Grey Parrot (*Psittacus erithacus*): Further Evidence for Comprehension of Categories and Labels," *Journal of Comparative Psychology*, 1990, *104*, 41–52; I. M. Pepperberg, "Referential Mapping: A Technique for Attaching Functional Significance to the Innovative Utterances of an African Grey Parrot" (*Psittacus erithacus*), *Applied Psycholinguistics*, 1990, *11*, 23–44; I. M. Pepperberg, "Conceptual Abilities of Some Nonprimate Species, with an Emphasis on an African Grey Parrot," in S. T. Parker and K. R. Gibson (Eds.), *"Language" and Intelligence in Monkeys and Apes* (New York: Cambridge University Press, 1990), pp. 469–507; I. M. Pepperberg, "A Communicative Approach to Animal Cognition: A Study of Conceptual Abilities of an African Grey Parrot," in C. A. Ristau (Ed.), *Cognitive Ethology: The Minds of Other Animals* (Hillsdale, N.J.: Lawrence Erlbaum Associates, 1991), pp. 153–186.

5. All of the data mentioned above are in Professor Pepperberg's original scientific papers (see note 4) but they are not emphasized in

these papers. Instead, the results emphasized are those that can be formulated in the more abstract and scientific-sounding phraseology of present-day psychology. For instance, in a representative paper Pepperberg writes:

> An African Grey parrot has been taught to use the sounds of English speech to identify, request, refuse, categorize, and quantify more than 80 different objects and to respond to questions concerning categorical concepts of color and shape. The parrot, Alex, has now been trained and tested on relational concepts of *same* and *different*. He learned to reply with the correct English categorical label ("color," "shape," or "mah-mah" [matter]) when asked "What's same?" or "What's different?" about pairs of objects that varied with respect to any combination of attributes. He performed equally well on pairs of novel and familiar objects, and special trials demonstrated that his responses were based upon the questions being posed as well as the attributes of the objects. (Pepperberg, "A Communicative Approach," p. 153 [4].)

A very striking aspect of Pepperberg's research is her very tight controls for bias, especially for the Clever Hans effect, or experimenter bias. To control for cueing by the experimenters, those who train Alex on a certain task never test him on that task. To control for expectational cueing, the topics he is asked about are intermingled; he never knows what he will be asked about next, and the order of both the test questions and the test objects is randomized. To be sure that Alex's responses are not stimulus specific, sometimes he is tested with entirely novel objects. Although various types of bias are common in behavioral research (paradigm bias, experimental design bias, data analysis bias, investigator bias, experimenter bias, etc.), Pepperberg's research with Alex excludes bias effects more rigorously than any of the many research projects I have studied intensively. T. X. Barber, *Pitfalls in Human Research: Ten Pivotal Points* (Elmsford, N.Y.: Pergamon Press, 1976).

6. R. J. Herrnstein, "Objects, Categories, and Discriminative Stimuli," in H. L. Roitblat, T. G. Bever, and H. S. Terrace (Eds.), *Animal Cognition* (Hillsdale, N.J.: Lawrence Erlbaum Associates, 1984), pp. 233–261; R. J. Herrnstein and D. H. Loveland, "Complex Visual Concepts in the Pigeon," *Science*, 1964, *146*, 549–551; R. J.

Herrnstein, D. H. Loveland, and C. Cable, "Natural Concepts in Pigeons," *Journal of Experimental Psychology: Animal Behavior Processes*, 1976, *2*, 285–302; J. Poole and D. G. Lander, "The Pigeon's Concept of Pigeon," *Psychonomic Science*, 1971, *25*, 157–158.

7. J. Cerella, "Visual Classes and Natural Categories in the Pigeon," *Journal of Experimental Psychology: Human Perception and Performance*, 1979, *5*, 68–77; J. Cerella, "The Pigeon's Analysis of Pictures," *Pattern Recognition*, 1980, *12*, 1–6; J. D. Delius and B. Nowak, "Visual Symmetry Recognition by Pigeons," *Psychological Research*, 1982, *44*, 199–212; R. E. Lubow, "Higher-order Concept Formation in the Pigeon," *Journal of the Experimental Analysis of Behavior*, 1974, *21*, 475–483.

The analysis of pigeon conceptual abilities, which began with the paper by Herrnstein and Loveland in 1964 (see note 6) is continuing. A relatively recent paper reported that, after nearly a quarter century of work in the area, "the conceptual abilities of pigeons are more advanced than hitherto suspected." R. S. Bhatt, E. A. Wasserman, W. F. Reynolds, Jr., and K. S. Knauss, "Conceptual Behavior in Pigeons: Categorization of Both Familiar and Novel Examples from Four Classes of Natural and Artifical Stimuli," *Journal of Experimental Psychology: Animal Behavior Processes*, 1988, *14*, 219–234.

Since the researchers in this area adhere to a Skinnerian behavioristic approach to psychology, they avoid implying that their captive pigeons' concept-forming abilities suggest intelligence or mentality. Nevertheless, the astonishing mental capabilities of pigeons shine through the "hard" Skinnerian data and reductionistic interpretations. For instance, Herrnstein discusses the findings as follows:

> . . . organisms devoid of language, and presumably also of the associated higher cognitive capacities, can rapidly extract abstract invariances from some (but not all) stimulus classes containing instances so variable that we cannot physically describe either the class rule or the instances, let alone account for the underlying capacity. . . . Animals that are celebrated more for their lack of intelligence than the reverse can sort exemplars of such variety that they out-perform the most ambitious computer simulations (or even the most

ambitious theories of a simulation). . . . Each new effort to push the limits of animals' conceptual abilities seems to have a good chance of finding new abilities. The limits to date are more in the experiments tried with animals than in the animals.

R. J. Herrnstein, "Riddles of Natural Categorization," in L. Weiskrantz (Ed.), *Animal Intelligence* (New York: Oxford University Press, 1985), pp. 129–144. (The quotation is from pages 129 and 133.)

8. Pigeons' ability to hold a sequence in mind is discussed by H. S. Terrace, "Animal Cognition," in Roitblat, Bever, and Terrace (Eds.), *Animal Cognition* pp. 7–28 [6]; R. Epstein, C. E. Kirshnit, R. P. Lanza, and L. C. Rubin, "Insight in the Pigeon: Antecedents and Determinants of Intelligent Performance," *Nature*, March 1, 1984, *308*, 61–62.

9. L. J. Stettner and K. A. Matyniak, "The Brain of Birds," in B. B. Wilson (Ed.), *Birds: Readings from Scientific American* (San Francisco: W. H. Freeman, 1980), pp. 192–199.

10. D. S. Blough, "Pigeon Perception of Letters of the Alphabet," *Science*, 1982, *218*, 397–398.

11. S. Brownlee, "Intelligence: A Riddle Wrapped in a Mystery," *Discover*, October 1985, *6* (No. 10), 85–93; M. Menne and E. Curio, "Investigations into the Symmetry Concept of the Great Tit (*Parus major*)," *Zeitschrift für Tierpsychologie*, 1978, *47*, 299–322; N. Pastore, "Learning in the Canary," *Scientific American*, June 1955, *192* (No. 6), 72–79; C. M. E. Ryan, "Concept Formation and Individual Recognition in the Domestic Chicken (*Gallus gallus*)," *Behavioral Analysis Letter*, 1982, *2*, 213–220; C. M. E. Ryan, "Mechanisms of Individual Recognition in Birds," Master's thesis, University of Exeter, 1982; W. H. Thorpe, "Animal Learning," *Encyclopedia Britannica* (15th edition), 1977, *10*, 731–746.

12. R. Baker, *The Mystery of Migration* (New York: Viking Press, 1981), p. 129; R. P. Balda, "Recovery of Cached Seeds by a Captive *Nucifraga caryocatactes*," *Zeitschrift für Tierpsychologie*, 1980, *52*, 331–346; R. P. Balda and R. J. Turek, "The Cache-Recovery System as an Example of Memory Capabilities in Clark's Nutcracker," in Roitblat, Bever, and Terrace (Eds.), *Animal Cognition* pp. 513–532 [6]; R. J. Cowie, J. R. Krebs, and D. F. Sherry,

"Food Storing in Marsh Tits," *Animal Behaviour*, 1981, *29*, 1252–1259; A. C. Kamil, "Adaptation and Cognition: Knowing What Comes Naturally," in Roitblat, Bever, and Terrace (Eds.), *Animal Cognition* pp. 533–544 [6]; S. J. Shettleworth and J. R. Krebs, "How Marsh Tits Find Their Hoards: The Role of Site Preference and Spatial Memory," *Journal of Experimental Psychology: Animal Behavior Processes*, 1982, *8*, 354–375; S. B. Vander Wall, "An Experimental Analysis of Cache Recovery in Clark's Nutcracker," *Animal Behaviour*, 1982, *30*, 84–94; S. B. Vander Wall and R. P. Balda, "Ecology and Evolution of Food-storing Behavior in Conifer-seed-caching Corvids," *Zeitschrift für Tierpsychologie*, 1981, *56*, 217–242.

13. B. B. Beck, *Animal Tool Behavior: The Use and Manufacture of Tools by Animals* (New York: Garland STPM Press, 1980), p. 22; A. J. Marshall, *Bower-Birds* (London: Oxford University Press, 1954).

14. D. Lack, *Darwin's Finches* (New York: Harper & Row, 1961).

15. M. Allaby, *Animal Artisans* (New York: Alfred A. Knopf, 1982), p. 301; D. C. Gayou, "Tool Use by Green Jays," *Wilson Bulletin*, 1982, *94*, 593–594; A. Hardy, *The Living Stream: Evolution and Man* (New York: Harper & Row, 1965), p. 176.

16. Beck, *Animal Tool Behavior*, p. 29 [13]; T. Jones and A. Kamil, "Tool-making and Tool-using in the Northern Blue Jay," *Science*, 1973, *180*, 1076–1078; W. W. Judd, "A Blue Jay in Captivity for Eighteen Years," *Bird Banding*, 1975, *46*, 250; J. B. Reid, "Tool-use by a Rook (*Corvus frugilegus*) and Its Causation," *Animal Behaviour*, 1982, *30*, 1212–1216.

17. J. R. Michener, "California Jays: Their Storage and Recovery of Food and Observations at One Nest," *Condor*, 1945, *47*, 206–210.

18. J. Bonnet, "Comportement curieux d'un Grand Corbeau à son site de nidification," *Alauda*, 1986, *54*, 71; S. Janes, "The Apparent Use of Rocks by a Raven in Nest Defense," *Condor*, 1976, *78*, 409.

19. T. Angell, *Ravens, Crows, Magpies, and Jays* (Seattle: University of Washington Press, 1978); Beck, *Animal Tool Behavior* [13]; D. B. Grobecker and T. W. Pietsch, "Crows' Use of Automobiles as Nutcrackers," *Auk*, 1978, *95*, 760–761; R. Zach, "Selection and Dropping of Whelks by Northwestern Crows," *Behaviour*, 1978, *67*, 134–148.

20. L. Homberg, Fiskande Kråkor, *Fauna och Flora*, 1957, *5*, 182–185.

21. Beck, *Animal Tool Behavior* [13]; H. Higuchi, "Bait-fishing by the Green-backed Heron *Ardeola striata* in Japan," *Ibis*, 1986, *128*,

285–290; H. Higuchi, "Cast Master," *Natural History*, 1987, *96* (No. 8), 40–43; H. B. Lovell, "Baiting of Fish by a Green Heron," *Wilson Bulletin*, 1958, *70*, 280–281; R. Sisson, "Aha! It Really Works!" *National Geographic Magazine*, 1974, *145*, 142–147.

22. D. R. Griffin, "Progress Toward a Cognitive Ethology," in Ristau (Ed.), *Cognitive Ethology: The Minds of Other Animals* pp. 3–17 [4].

23. Thorpe, "Animal Learning" [11]; J. van Lawick-Goodall, "Tool-using in Primates and Other Vertebrates," in D. Lehrman, R. Hinde, and E. Shaw (Eds.), *Advances in the Study of Behavior*, Vol. 3 (New York: Academic Press, 1970), pp. 195–249.

24. Allaby, *Animal Artisans* pp. 9, 30 [15]; P. A. DeBenedictis, "The Bill-brace Feeding Behavior of the Galapagos Finch *Geospiza conirostris*," *Condor*, 1966, *68*, 206–208; C. H. Fry, "The Biology of African Bee-eaters," *Living Bird*, 1972, *11*, 75–112; G. J. Romanes, *Animal Intelligence* (New York: D. Appleton & Co., 1883), p. 283.

   There are many other reports of tool use in birds. A useful guide to the rich literature is presented by I. M. Pepperberg, "Tool Use in Birds: An Avian Monkey Wrench," *Behavioral and Brain Sciences*, 1989, *12*, 604–605.

25. Angell, *Ravens, Crows, Magpies, and Jays* [19]; C. Darwin, "A Posthumous Essay on Instinct," in G. J. Romanes, *Mental Evolution in Animals with a Posthumous Essay on Instinct by Charles Darwin* (London: Kegan Paul, Trench, 1883), pp. 355–384; D. Lack, *The Life of the Robin* (London: Penguin Books, 1953); R. F. Leslie, *Lorenzo the Magnificent: The Story of an Orphaned Blue Jay* (New York: W. W. Norton, 1985); Romanes, *Animal Intelligence* [24]; A. F. Skutch, *Parent Birds and Their Young* (Austin: University of Texas Press, 1976); M. B. Trautman, "Courtship Behavior of the Black Duck," *Wilson Bulletin*, 1947, *59*, 26–35.

26. J. D. Ligon and S. H. Ligon, "Communal Breeding in Green Woodhoopoes," *Nature*, 1978, *276*, 496–498; Romanes, *Animal Intelligence* [24]; A. F. Skutch, *Helpers at Birds' Nests* (Iowa City: University of Iowa Press, 1987).

Chapter 2. Avian Flexibility

1. J. C. Welty and L. Baptista, *The Life of Birds* (4th Ed.) (New York: Saunders, 1988), p. 75.

2. F. B. Gill, *Ornithology* (New York: W. H. Freeman, 1990), p. 277.

3. F. B. Gill and L. L. Wolf, "Foraging Strategies and Energetics of East African Sunbirds at Mistletoe Flowers," *American Naturalist*, 1975, *109*, 491–510; L. Howard, *Birds as Individuals* (London: Readers Union, Collins, 1953); Welty and Baptista, *The Life of Birds*, pp. 248–251 [1].

4. J. R. Krebs, M. H. MacRoberts, and J. M. Cullen, "Flocking and Feeding in the Great Tit (*Parus major*)—An Experimental Study," *Ibis*, 1972, *114*, 507–530.

5. C. Ogburn, *The Adventure of Birds* (New York: William Morrow, 1976), pp. 107–108; D. A. Vleugel, "A Case of Herring Gulls Learning by Experience to Feed After Explosions by Mines," *British Birds*, 1951, *44*, 180.

6. A. H. Verrill, *Strange Birds and Their Stories* (New York: Page, 1938); Welty and Baptista, *The Life of Birds*, pp. 115–116 [1].

7. J. Fisher and R. A. Hinde, "The Opening of Milk Bottles by Birds," *British Birds*, 1949, *42*, 347–357; R. A. Hinde and J. Fisher, "Further Observations on the Opening of Milk Bottles by Birds," *British Birds*, 1951, *44*, 393–396; J. C. Welty, *The Life of Birds* (New York: Alfred A. Knopf, 1963), p. 168.

8. A. F. Skutch, *Parent Birds and Their Young* (Austin: University of Texas Press, 1976), pp. 261–262, 277.

9. A. Keast and E. S. Morton (Eds.), *Migrant Birds in the Neotropics: Ecology, Behavior, Distribution, and Conservation* (Washington, D.C.: Smithsonian Institution Press, 1980); R. F. Pasquier and E. S. Morton, "For Avian Migrants a Tropical Vacation Is Not a Bed of Roses," *Smithsonian*, October 1982, *13* (No. 7), 169–188.

10. C. Darwin, "A Posthumous Essay on Instinct," in G. J. Romanes, *Mental Evolution in Animals with a Posthumous Essay on Instinct by Charles Darwin* (London: Kegan Paul, Trench, 1883), pp. 355–384.

11. G. J. Romanes, *Animal Intelligence* (New York: D. Appleton & Co., 1883); Welty and Baptista, *The Life of Birds*, p. 290 [1].

12. Verrill, *Strange Birds and Their Stories*, p. 191 [6].

13. E. von Hartmann, *Philosophie des Unbewussten* (3d Ed.) (Berlin: Duncker, 1871); Welty and Baptista, *The Life of Birds*, p. 295 [1].

14. N. E. Collias and E. C. Collias, *Nest Building and Bird Behavior* (Princeton, N.J.: Princeton University Press, 1984).

15. R. P. Balda and R. J. Turek, "The Cache-recovery System as an Example of Memory Capabilities in Clark's Nutcracker," in H. L. Roitblat, T. G. Bever, and H. S. Terrace (Eds.), *Animal Cognition*

(Hillsdale, N.J.: Lawrence Erlbaum Associates, 1984), pp. 513–532.

16. J. G. Myers, "Nesting Association of Birds with Social Insects," *Transactions of the Entomological Society of London*, 1935, *83*, 11–22; Skutch, *Parent Birds and Their Young* [8]; Welty and Baptista, *The Life of Birds*, p. 295 [1].

17. D. R. Griffin, *Animal Thinking*, (Cambridge, Mass.: Harvard University Press, 1984), pp. 107–110.

18. Gill, *Ornithology*, pp. 376–377 [2]; Romanes, *Animal Intelligence*, p. 289 [11]; Welty, *The Life of Birds*, p. 336 [7].

19. M. Bright, *Animal Language* (Ithaca, N.Y.: Cornell University Press, 1984), p. 126; E. Curio, "Cultural Transmission of Enemy Recognition," *Science*, 1978, *202*, 899–901; R. Drent, "Incubation," in D. S. Farner and J. R. King (Eds.), *Avian Biology*, Vol. 5 (New York: Academic Press, 1975), pp. 333–420.

20. Griffin, *Animal Thinking*, [17]; C. A. Ristau, "Aspects of the Cognitive Ethology of an Injury-feigning Bird, the Piping Plover," in C. A. Ristau (Ed.), *Cognitive Ethology: The Minds of Other Animals* (Hillsdale, N.J.: Lawrence Erlbaum Associates, 1991), pp. 91–126; Skutch, *Parent Birds and Their Young*, pp. 414–415 [8].

21. Ristau, "Aspects of the Cognitive Ethology of an Injury-feigning Bird" [20].

22. Griffin, *Animal Thinking*, pp. 87–94 [17]; Skutch, *Parent Birds and Their Young*, pp. 414–415 [8].

23. T. Angell, *Ravens, Crows, Magpies, and Jays* (Seattle: University of Washington Press, 1978), p. 61; L. Kilham, *The American Crow and the Common Raven* (College Station: Texas A & M University Press, 1989), pp. 37, 113–115, 119–120; R. F. Pasquier, *Watching Birds: An Introduction to Ornithology* (Boston: Houghton Mifflin, 1977), p. 100; Skutch, *Parent Birds and Their Young*, pp. 333–336 [8].

24. P. de Groot, "Information Transfer in Socially Roosting Weaver Birds (*Quelea quelea: Ploceinae*): An Experimental Study," *Animal Behaviour*, 1980, *28*, 1249–1254.

25. Skutch, *Parent Birds and Their Young* [8].

26. Howard, *Birds as Individuals* [3]; T. Mebs, "Family: Falcons," in B. Grizmek (Ed.), *Grizmek's Animal Life Encyclopedia*, Vol. 7 (New York: Van Nostrand Reinhold, 1972); Welty, *The Life of Birds*, p. 183 [7]; T. M. Caro and M. D. Hauser, "Is There Teaching in Nonhuman Animals? *Quarterly Review of Biology*, 1992, *67*, 13–174.

27. Many examples from the earlier literature are presented in C.

Darwin, *The Descent of Man and Selection in Relation to Sex* (London: John Murray, 1871). See also Chapter 8 herein.

28. Balda and Turek, "The Cache-recovery System" [15].

29. H. N. Southern, "The Natural Control of a Population of Tawny Owls (*Strix aluco*)," *Journal of Zoology*, 1970, *162*, 197–285.

30. D. Lack, *The Natural Regulation of Animal Numbers* (Oxford: Clarendon Press, 1954); J. D. Ligon, "Reproductive Interdependence of Piñon Jays and Piñon Pines," *Ecological Mogographs*, 1978, *48*, 111–126.

Chapter 3. Instincts Guide Both Birds and Humans

1. P. D. Eimas and V. C. Tartter, "On the Development of Speech Perception: Mechanisms and Analogies," in H. W. Reese and L. P. Lipsitt (Eds.), *Advances in Child Development and Behavior*, Vol. 13 (New York: Academic Press, 1979); P. D. Eimas, "Speech Perception: A View of the Initial State and Perceptual Mechanisms," in J. Mehler, E. Walker, and M. Garrett (Eds.) *Perspectives in Mental Representation* (Hillsdale, N. J.: Lawrence Erlbaum Associates, 1982); P. D. Eimas, "Constraints on a Model of Infant Speech Perception," in J. Mehler and R. Fox (Eds.), *Neonate Cognition: Beyond the Blooming Buzzing Confusion* (Hillsdale, N.J.: Lawrence Erlbaum Associates, 1985), pp. 185–197; P. K. Kuhl, "Categorization of Speech by Infants," in Mehler and Fox (Eds.), *Neonate Cognition: Beyond the Blooming Buzzing Confusion*, pp. 231–262.

2. The material in this and subsequent paragraphs on the human speech-language instinct is based primarily on the following: N. Chomsky, "A Review of B. F. Skinner's *Verbal Behavior*," *Language*, 1959, *35*, 26–58; N. Chomsky, "A Transformational Approach to Syntax," in J. A. Fodor and J. J. Katz (Eds.), *The Structure of Language: Readings in the Philosophy of Language* (Englewood Cliffs, N.J.: Prentice-Hall, 1964), pp. 211–245; N. Chomsky, *Language and Mind* (New York: Harcourt Brace Jovanovich, 1972); N. Chomsky, *Reflections on Language* (New York: Pantheon, 1975); N. Chomsky, *Rules and Representations* (New York: Columbia University Press, 1980); H. Gardner, *The Mind's New Science: A History of the Cognitive Revolution* (New York: Basic Books, 1985), pp. 182–222; E. H. Lenneberg, *Biological Foundations of Language* (New

York: John Wiley, 1967); P. Lieberman, *The Biology and Evolution of Language* (Cambridge, Mass.: Harvard University Press, 1984); M. Piattelli-Palmarini, *Language and Learning: The Debate Between Jean Piaget and Noam Chomsky* (Cambridge, Mass.: Harvard University Press, 1980)

3. P. Marler, "Bird Song and Speech Development: Could There be Parallels?" *American Scientist*, 1970, *58*, 669–673; P. Marler, "Song Learning: Innate Species Differences in the Learning Process," in P. Marler and H. S. Terrace (Eds.), *The Biology of Learning* (New York: Springer-Verlag, 1984), pp. 289–309.

4. F. Nottebohm, "Vocal Behavior in Birds," in D. S. Farner and J. R. King (Eds.), *Avian Biology*, Vol. 5 (New York: Academic Press, 1975), pp. 287–332; G. Montgomery, "A Brain Reborn," *Discover*, June, 1990, 48–53.

5. Research on the instincts of humans has been most recently summarized (under the term phylogenetic adaptations) in I. Eibl-Eibesfeldt, *Human Ethology* (Hawthorn, N.Y.: Aldine de Gruyter, 1989). Also useful are: T. G. R. Bower, *Development in Infancy* (San Francisco: W. H. Freeman, 1974); C. Buhler, P. Keith-Spiegel, and K. Thomas, "Developmental Psychology," in B. B. Wolman (Ed.), *Handbook of General Psychology* (Englewood Cliffs, N.J.: Prentice-Hall, 1973), pp. 861–917; W. James, *The Principles of Psychology*, Vol. 2, (New York: Dover, 1950), pp. 383–441; J. Kagan, "Development of Human Behavior," *Encyclopedia Britannica*, 15th ed., 1977, *8*, 1136–1146; Mehler and Fox (Eds.), *Neonate Cognition: Beyond the Blooming Buzzing Confusion* [1]; N. Munn, *Psychological Development* (Boston: Houghton Mifflin, 1938); M. Roberts, *Infant Crying* (New York: Plenum, 1985); W. H. Thorpe, *Animal Nature and Human Nature* (New York: Doubleday, 1974); W. H. Thorpe, "Animal Learning," *Encyclopedia Britannica*, 15th ed., 1977, *10*, 731–746; R. J. Trotter, "You've Come a Long Way, Baby," *Psychology Today*, May 1987, 21 (No. 5), 34–45.

6. T. M. Field, R. Woodson, R. Greenberg, and D. Cohen, "Discrimination and Imitation of Facial Expressions by Neonates," *Science*, 1982, *218*, 179–181; A. N. Meltzoff and M. K. Moore, "Imitation of Facial and Manual Gestures by Human Neonates," *Science*, 1977, *198*, 75–78; A. N. Meltzoff and M. K. Moore, "Newborn Infants Imitate Adult Facial Gestures," *Child Development*, 1983, *54*, 702–709; A. N. Meltzoff and M. K. Moore, "Cognitive Foundations and Social Functions of Imitation and Intermodal Representation in

Infancy," in Mehler and Fox (Eds.), *Neonate Cognition*, pp. 139–156 [5]. A brief critique of this research, which concludes that "young babies do sometimes *appear* to imitate facial gestures," is presented in D. Maurer and C. Maurer, *The World of the Newborn* (New York: Basic Books, 1988), pp. 274–275. Another recent critique concludes that the newborn's precocious imitation abilities are "species-typical neonatal sucking and grasping 'reflexes.' " S. Chevalier-Skolnikoff, "Tool Use in *Cebus*: Its Relation to Object Manipulation, the Brain, and Ecological Adaptations," *Behavioral and Brain Sciences*, 1989, *12*, 610–621.

7. I. Wyllie, *The Cuckoo* (New York: Universe Books, 1981); H. Friedmann, "Cuculiforms," *Encyclopedia Britannica*, 15th ed. 1977, *5*, 358–361.

8. Background material leading to the very important conclusion that *instinctual guidelines for living have to be implemented flexibly to fit varying circumstances* is found in: J. L. Gould, *Ethology* (New York: W. W. Norton, 1982); J. L. Gould and C. G. Gould, *The Honey Bee* (New York: Scientific American Library, 1988); J. L. Gould and P. Marler, "Learning by Instinct," *Scientific American*, January 1987, *256* (No. 1), 74–85; D. R. Griffin, *Animal Thinking* (Cambridge, Mass.: Harvard University Press, 1984); R. W. G. Hingston, *Instinct and Intelligence* (New York: Macmillan, 1929); A. Koestler, *The Act of Creation* (New York: Macmillan, 1964); A. Koestler, *Janus: A Summing Up* (New York: Random House, 1978); G. A. Miller, E. Galanter, and K. H. Pribram, *Plans and the Structure of Behavior* (New York: Henry Holt, 1960); Thorpe, *Animal Nature* [5]; Thorpe, "Animal Learning" [5]; J. Z. Young, *Programs of the Brain* (New York: Oxford University Press, 1978).

9. Additional research, summarized by Gould and Marler (see note 8), leads to the important conclusion that each kind of animal is instinctually programmed to learn some things quickly and easily and other things with great difficulty or not at all. For example, as we will discuss in Chapter 7, birds such as the Arctic tern are as instinctually programmed to learn to navigate by interpreting and synthesizing cues in nature as humans are instinctually programmed to learn to speak by interpreting and synthesizing sounds in their environment. Also, as Hingston emphasized some years ago (see note 8), instincts themselves have the hallmarks of intel-

ligence. Paradoxically, intelligence underlies instincts, and there is an instinctual basis for intelligence.

Chapter 4. Avian Languages

1. M. Argyle, *Bodily Communication* (New York: International Universities Press, 1975); I. Eibl-Eibesfeldt, *Human Ethology* (New York: Aldine de Gruyter, 1989).

2. F. B. Gill, *Ornithology* (New York: W. H. Freeman, 1990), pp. 173–178; R. F. Leslie, *Lorenzo the Magnificent: The Story of an Orphaned Blue Jay* (New York: W. W. Norton, 1985); W. J. Smith, "Communication in Birds," in T. A. Sebeok (Ed.), *How Animals Communicate* (Bloomington: Indiana University Press, 1977), pp. 545–574.

3. L. Howard, *Birds as Individuals* (London: Readers Union, Collins, 1953), p. 143.

4. Ibid., p. 146.

5. Leslie, *Lorenzo the Magnificent* [2].

6. Gill, *Ornithology* [2]; Smith, "Communication in Birds" [2].

7. P. de Groot, "Information Transfer in Socially Roosting Weaver Bird (*Quelea quelea: Ploceinae*): An Experimental Study," *Animal Behaviour*, 1980, *28*, 1249–1254; Gill, *Ornithology*, pp. 173–178 [2]; F. Nottebohm, "Vocal Behavior in Birds," in D. S. Farner and J. R. King (Eds.), *Avian Biology*, Vol. 5 (New York: Academic Press, 1975), pp. 287–332; J. H. Prince, *Languages of the Animal World* (Nashville, Tenn.: Thomas Nelson, 1975); A. F. Skutch, *Parent Birds and Their Young* (Austin: University of Texas Press, 1976).

8. H. Frings and M. Frings, "The Language of Crows," *Scientific American*, November, 1959, *201* (No. 5), 119–131; B. Gilbert, *How Animals Communicate* (New York: Pantheon, 1966); C. Ogburn, *The Adventure of Birds* (New York: William Morrow, 1976).

9. R. Jellis, *Bird Sounds and Their Meaning* (London: British Broadcasting Corporation, 1977); M. Konishi, "Time Resolution by Single Auditory Neurons in Birds," *Nature*, 1969, *222*, 566–567; J. C. Welty and L. Baptista, *The Life of Birds* (4th Ed.) (New York: Saunders, 1988), p. 82.

10. Howard, *Birds as Individuals*, p. 147 [3]. The critical differences in

the avian and human experience of time are also discussed by L. J. Halle, *The Appreciation of Birds* (Baltimore: Johns Hopkins University Press, 1989), p. 95.

11. C. G. Beer, "Multiple Functions and Gull Displays," in G. Baerends, C. G. Beer, and A. Manning (Eds.), *Essays on Function and Evolution in Behavior: A Festschrift for Professor Niko Tinbergen* (Oxford: The Clarendon Press, 1975), chap. 2.

12. J. LeComte and D. Koechlin-Schwartz, *How to Talk to the Birds and the Beasts* (New York: Arbor House, 1980), pp. 114–117.

13. J. T. Bonner, *The Evolution of Culture in Animals* (Princeton, N.J.: Princeton University Press, 1980), p. 112.

14. A. P. Moller, "False Alarm Calls as a Means of Resource Usurpation in the Great Tit *Parus major*," *Ethology*, 1988, *79*, 25–30; C. A. Munn, "Birds that 'Cry Wolf,' " *Nature*, 1986, *319*, 143–145; C. A. Munn, "The Deceptive Use of Alarm Calls by Sentinel Species in Mixed-species Flocks of Neotropical Birds," in R. W. Mitchell and N. S. Thompson (Eds.), *Deception: Perspectives on Human and Nonhuman Deceit* (Albany, N.Y.: SUNY Press, 1986).

15. Frings and Frings, "The Language of Crows" [8].

16. C. Simonds, *Private Life of Garden Birds* (Emmaus, Pa.: Rodale Press, 1984), p. 41.

17. B. Heinrich, *One Man's Owl* (Princeton, N.J.: Princeton University Press, 1987), p. 126.

18. Leslie, *Lorenzo the Magnificent* [2]; observations by Frank M. Chapman are presented in Ogburn, *The Adventure of Birds* [8]; Prince, *Languages of the Animal World*, p. 75 [7].

19. T. Angell, *Ravens, Crows, Magpies, and Jays* (Seattle: University of Washington Press, 1978).

20. Prince, *Languages of the Animal World*, p. 75 [7].

21. C. G. Beer, "Some Complexities in the Communication Behavior of Gulls," in S. R. Harnad, H. D. Steklis, and J. Lancaster (Eds.), *Origin and Evolution of Language and Speech* (Annals of the New York Academy of Sciences, Vol. 280) (New York: New York Academy of Sciences, 1976), pp. 413–432.

Chapter 5. Lorenzo, the Communicative Jay

1. Representative examples of different types of modern ornithology texts include: R. Burton, *Bird Behavior* (New York: Alfred A.

Knopf, 1985); F. B. Gill, *Ornithology* (New York: W. H. Freeman, 1990); J. C. Welty and L. Baptista, *The Life of Birds* (4th Ed.) (New York: W. B. Saunders, 1988).

2. R. F. Leslie, *Lorenzo the Magnificent: The Story of an Orphaned Blue Jay* (New York: W. W. Norton, 1985).

3. Ibid., p. 93.

Chapter 6. Avian Music, Crafts, and Play

1. A. Skutch, "Bird Song and Philosophy," in L. E. Hahn (Ed.), *The Philosophy of Charles Hartshorne* (La Salle, Ill.: Open Court, 1991), pp. 65–76. The quotation is from page 69.

2. M. Fox, *The Coming of the Cosmic Christ* (San Francisco: Harper & Row, 1988), p. 14.

3. L. Howard, *Birds as Individuals* (London: Reader's Union, Collins, 1953).

4. Ibid., pp. 184–185.

5. C. Hartshorne, *Born to Sing: An Interpretation and World Survey of Bird Song* (Bloomington: Indiana University Press, 1973).

6. Skutch, "Bird Song and Philosophy" [1].

7. C. G. Beer, "Study of Vertebrate Communication—Its Cognitive Implications," in D. R. Griffin (Ed.), *Animal Mind–Human Mind* (New York: Springer-Verlag, 1982), pp. 251–267.

8. Hartshorne, *Born to Sing*, pp. 107–109 [5]; R. Jellis, *Bird Sounds and Their Meaning* (London: British Broadcasting Corporation, 1977); M. Konishi, "Time Resolution by Single Auditory Neurons in Birds," *Nature*, 1969, *222*, 566–567; P. Marler, "Song Learning: Innate Species Differences in the Learning Process," in P. Marler and H. S. Terrace (Eds.), *The Biology of Learning* (New York: Springer-Verlag, 1984), pp. 289–309; J. C. Welty and L. Baptista, *The Life of Birds* (4th Ed.) (New York: W. B. Saunders, 1988), pp. 82, 224.

9. Hartshorne, *Born to Sing*, pp. 95–96 [5].

10. Ibid., p. 41.

11. C. K. Catchpole, "Sexual Selection and the Evolution of Complex Songs Among European Warblers of the Genus *Acrocephalus*," *Behaviour*, 1980, *74*, 149–166; C. K. Catchpole, B. Leisler, and H. Winkler, "Polygyny in the Great Reed Warbler, *Acrocephalus arun-*

∂inaceouð," *Behavioral Ecology and Sociobiology*, 1985, *16*, 285–291; J. B. Falls, "Territorial Song in the White-throated Sparrow," in R. A. Hinde (Ed.), *Bird Vocalizations* (Cambridge, England: Cambridge University Press, 1969); P. McGregor, J. Krebs, and C. Perrins, "Song Repertoires and Lifetime Reproductive Success in the Great Tit (*Parus major*)," *American Naturalist*, 1981, *118*, 149–159.

12. D. E. Kroodsma, "Correlates of Song Organization Among North American Wrens," *American Naturalist*, 1977, *3*, 995–1008; D. E. Kroodsma, "Winter Wren Singing Behavior: A Pinnacle of Song Complexity," *Condor*, 1980, *82*, 357–365; M. M. Nice, "Studies in the Life History of the Song Sparrow," *Transactions of the Linnaean Society of New York*, 1943, 6, 1–328; W. J. Smith, "Communication in Birds," in T. A. Sebeok (Ed.), *How Animals Communicate* (Bloomington: Indiana University Press, 1977), pp. 545–574; Welty and Baptista, *The Life of Birds*, p. 214 [7].

13. Howard, *Birds as Individuals*, p. 178 [3].

14. Ibid., p. 178–179.

15. Ibid., pp. 172–173.

16. C. Hartshorne, "The Relation of Bird Song to Music," *Ibis*, 1958, *100*, 421–445.

17. Howard, *Birds as Individuals*, p. 209 [3].

18. Ibid., p. 203.

19. Ibid., pp. 210–211.

20. Ibid., pp. 212–213.

21. W. H. Thorpe, "Duetting and Antiphonal Singing in Birds: Its Extent and Significance," *Behaviour: Monograph Supplement*, 1972, *18*, 1–197; W. H. Thorpe, "Duet-singing Birds," *Scientific American*, August 1973, 229 (No. 2), 70–79.

22. Jellis, *Bird Sounds* [8]; P. Kunkel, "Mating Systems of Tropical Birds," *Zeitschrift für Tierpsychologie*, 1974, *34*, 265–307; Thorpe, "Duetting" [21].

23. H. Gardner, *Frames of Mind: The Theory of Multiple Intelligences* (New York: Basic Books, 1983).

24. C. Darwin, *The Descent of Man and Selection in Relation to Sex* (London: John Murray, 1871).

25. J. D. Delius and G. Habers, "Symmetry: Can Pigeons Conceptualize It?" *Behavioral Biology*, 1978, *22*, 336–342; M. Menne and E. Curio, "Investigations into the Symmetry Concept of the Great Tit

(*Parus major*)," *Zeitschrift für Tierpsychologie*, 1978, *47*, 299–322; B. Rensch, "Die Wirksamkeit ästhetischer Factoren bei Wirbeltieren," *Zeitschrift für Tierpsychologie*, 1958, *15*, 447–461; M. Tigges, "Muster und Farbbevorzugung bei Fischen und Vögeln," *Zeitschrift für Tierpsychologie*, 1963, *20*, 129–142.

26. K. von Frisch, *Animal Architecture* (New York: Harcourt Brace Jovanovich, 1971), pp. 237–247.

27. J. Diamond, "The Bower Builders," *Discover*, June 1984, *5* (No. 6), 52–58.

28. Heinz Sielmann, quoted by von Frisch, *Animal Architecture*, pp. 243–244 [26].

29. N. E. Collias and E. C. Collias, *Nest Building and Bird Behavior* (Princeton, N. J.: Princeton University Press, 1984); A. F. Skutch, *Parent Birds and Their Young* (Austin: University of Texas Press, 1976); von Frisch, *Animal Architecture* [26].

30. L. Bertin and M. Burton, "Birds (Class Aves)," in M. Burton (Ed.), *The New Larousse Encyclopedia of Animal Life* (Rev. Ed.) (London: Paul Hamlyn, 1980), pp. 331–467 (the discussion on crows' nests is on pages 347–349); Collias and Collias, *Nest Building* [29]; L. Kilham, *The American Crow and the Common Raven* (College Station: Texas A & M University Press, 1989), pp. 65–68.

31. The lengthy period of practice that is required has been documented for the village weaver by Collias and Collias, *Nest Building*, pp. 211–214 [29].

32. von Frisch, *Animal Architecture*, pp. 204–207, 227 [26].

33. Bertin and Burton, "Birds (Class Aves)," p. 440 [30]; C. Ogburn, *The Adventure of Birds* (New York: William Morrow, 1976).

34. von Frisch, *Animal Architecture* [26]; Welty and Baptista, *The Life of Birds*, pp. 278–280, 302 [8].

35. N. Chomsky, *Reflections on Language* (New York: Random House, 1975); Skutch, *Parent Birds and Their Young*, p. 122 [29].

36. T. Angell, *Ravens, Crows, Magpies, and Jays* (Seattle: University of Washington Press, 1978); C. C. Bradley, "Play Behavior in Northern Ravens," *Passenger Pigeon*, 1978, *40*, 493–495; H. E. Burtt, *The Psychology of Birds: An Interpretation of Bird Behavior* (New York: Macmillan, 1967); E. Gwinner, "Ueber einige Bewegungsspiele des Kolkraben (*Corvus corax* L.)," *Zeitschrift für Tierpsychologie*, 1966, *23*, 28–36; F. H. Herrick, "The Daily Life of the American Eagle: Late Phase," *Auk*, 1924, *41*, 517–541; B. King, "Hooded

Crows Dropping and Transferring Objects from Bill to Foot in Flight," *British Birds*, 1969, *62*, 201; J. Nicolai, "Mimicry in Parasitic Birds," in B. W. Wilson (Ed.), *Birds: Readings from Scientific American* (San Francisco: W. H. Freeman, 1980), pp. 135–141; W. H. Thorpe, *Learning and Instinct in Animals* (Cambridge, Mass.: Harvard University Press, 1956), p. 323; Welty and Baptista, *The Life of Birds*, pp. 188–189 [8]; E. O. Wilson, *Sociobiology: The New Synthesis* (Cambridge, Mass.: Harvard University Press, 1975), p. 166.

37. Howard, *Birds as Individuals* [3].
38. E. A. Armstrong, *Bird Display* (Cambridge, England: Cambridge University Press, 1942); A. H. Verrill, *Strange Birds and Their Stories* (New York: Page, 1938); Welty and Baptista, *The Life of Birds*, pp. 273–276 [8].

Chapter 7. Avian Navigation

1. The data that led to the new paradigm of avian navigation via information integration are reviewed in the following publications: K. P. Able, "Mechanisms of Orientation, Navigation, and Homing," in S. A. Gauthreaux, Jr. (Ed.), *Animal Migration, Orientation, and Navigation* (New York: Academic Press, 1980), pp. 283–373; K. P. Able and V. P. Bingham, "The Development of Orientation and Navigation Behavior in Birds," *Quarterly Review of Biology*, 1987, *62*, 1–29; T. Alerstam, "The Course and Timing of Bird Migration," in D. J. Aidley (Ed.), *Animal Migration* (New York: Cambridge University Press, 1981), pp. 9–54; R. R. Baker, *Bird Navigation: The Solution of a Mystery?* (New York: Holmes & Meier, 1984); P. Berthold, "Migration: Control and Metabolic Physiology," in D. S. Farner and J. R. King (Eds.), *Avian Biology*, Vol. 5 (New York: Academic Press, 1975), pp. 77–128; S. T. Emlen, "Migration: Orientation and Navigation," in Farner and King (Eds.), *Avian Biology*, Vol. 5, pp. 129–219; E. C. Gerrard, *Instinctive Navigation of Birds* (Skye, Scotland: The Scottish Research Group, 1981); D. R. Griffin, *Bird Migration* (New York: Dover, 1974): W. T. Keeton, "The Orientation and Navigation of Birds," in Aidley (Ed.), *Animal Migration* pp. 81–104; F. Papi and H. G. Wallraff (Eds.), *Avian Navigation* (New York: Springer-Verlag,

1982); K. Schmidt-Koenig, *Avian Orientation and Navigation* (New York: Academic Press, 1979); K. Schmidt-Koenig and W. T. Keeton (Eds.), *Animal Migration, Navigation, and Homing* (New York: Springer-Verlag, 1978); C. Walcott and A. J. Lednor, *Bird Navigation,* in A. H. Brush and G. H. Clark, Jr. (Eds.), *Perspectives in Ornithology* (New York: Cambridge University Press, 1983), pp. 513–542.

2. R. R. Baker (Ed.), *The Mystery of Migration* (New York: Viking, 1981).

3. Emlen, "Migration," p. 148 [1].

4. Walcott and Lednor, *Bird Navigation,* p. 514 [1].

5. Emlen, "Migration" [1].

6. K. Hoffmann, "Versuche zu der im Richtungsfinden der Vögel enthaltenen Zeitschätzung," *Zeitschrift für Tierpsychologie,* 1954, *11,* 453–475; Keeton, "The Orientation and Navigation of Birds" [1]; G. Kramer, "Eine neue Methode zur Erforschung der Zugorientierung und die bisher damit erzielten Ergebnisse," *Proceedings of the 10th International Ornithological Congress,* Uppsala, 1951, pp. 269–280; G. Kramer, "Die Sonnenorientierung der Vögel," *Verhandlungen der Deutschen zoologische Gesellschaft,* 1953, 72–84; G. V. T. Matthews, "An Investigation of Homing Ability in Two Species of Gulls," *Ibis,* 1952, *94,* 243–264; G. V. T. Matthews, "Navigation in the Manx Shearwater," *Journal of Experimental Biology,* 1953, *30,* 370–396; Schmidt-Koenig and Keeton, *Animal Migration* [1].

7. J. L. Gould and C. G. Gould, *The Honey Bee* (New York: Scientific American Library, 1988), pp. 128–138.

8. Kramer, "Eine neue Methode" [6]; Kramer, "Die Sonnenorientierung" [6].

9. Hoffmann, "Versuche zu der im Richtungsfinden" [6].

10. E. G. F. Sauer, "Die Sternenorientierung nächtlich Ziehender Grasmücken (*Sylvia atricapilla, borin* und *curruca*)," *Zeitschrift für Tierpsychologie,* 1957, *14,* 29–70.

11. The nearly two dozen experiments in this series are listed in Emlen, "Migration," p. 163 [1].

12. S. T. Emlen, "Migratory Orientation in the Indigo Bunting, *Passerina cyanea.* Part I. Evidence for Use of Celestial Cues," *Auk,* 1967, *84,* 309–342; S. T. Emlen, "Migratory Orientation in the Indigo Bunting, *Passerina cyanea.* Part II. Mechanism of Celestial Orientation," *Auk,* 1967, *84,* 463–489; S. T. Emlen, "The Develop-

ment of Migratory Orientation in Young Indigo Buntings," *Living Bird*, 1969, *8*, 113–126; S. T. Emlen, "Celestial Rotation: Its Importance to the Development of Migratory Orientation," *Science*, 1970, *170*, 1198–1201; S. T. Emlen, "The Ontogenetic Development of Orientation Capabilities," in S. R. Galler, K. Schmidt-Koenig, G. J. Jacobs, and R. E. Belleville (Eds.), *Animal Orientation and Navigation* (Washington, D.C.: National Aeronautic and Space Administration, 1972), pp. 191–210; Emlen, "Migration" [1].

13. Alerstam, "The Course and Timing" [1].

14. M. L. Kreithen and W. T. Keeton, "Detection of Changes in Atmospheric Pressure by the Homing Pigeon, *Columba livia*," *Journal of Comparative Physiology*, 1974, *89*, 73–82.

15. T. C. Williams and J. M. Williams, "Orientation of Transatlantic Migrants," in Schmidt-Koenig and Keeton (Eds.), *Animal Migration*, pp. 239–251 [1].

16. Alerstam, "The Course and Timing," p. 50 [1].

17. Baker, *Bird Navigation*, pp. 56–63 [1]; Emlen, "Migration," pp. 148–151 [1].

18. Emlen, "Migration" [1].

19. B. Bruderer, "Do Migrating Birds Fly Along Straight Lines?" in Papi and Wallraff (Eds.), *Avian Navigation*, pp. 3–14 [1].

20. Emlen, "Migration," p. 149 [1].

21. D. E. Carr, *The Forgotten Senses* (Garden City, N.Y.: Doubleday, 1972); J. T. Erichsen, "Vision," in B. Campbell and E. Lack (Eds.), *A Dictionary of Birds* (London: British Ornithologists Union, 1985), pp. 623–629; J. D. Pettigrew, "A Role for the Avian Pecten Oculi in Orientation to the Sun," in Schmidt-Koenig and Keeton (Eds.), *Animal Migration*, pp. 42–54 [1]; J. A. Waldvogel, "The Bird's Eye View," *American Scientist*, 1990, *78*, 342–353; G. L. Walls, *The Vertebrate Eye and Its Adaptive Radiation* (Bloomfield Hills, Mo.: Cranbrook Institute of Science, 1942).

22. The literature on geomagnetic sensitivity in animals is summarized in T. H. Waterman, *Animal Navigation* (New York: Scientific American Library, 1989). The data on the geomagnetic sensitivity of humans is presented in: R. R. Baker, "Goal Orientation by Blindfolded Humans after Long-distance Displacement: Possible Involvement of a Magnetic Sense," *Science*, 1980, *210*, 555–557; R. R. Baker, "Man and Other Vertebrates: A Common Perspective to Migration and Navigation," in Aidley (Ed.), *Animal Migra-*

tion, pp. 241–260 [1]; R. R. Baker, *Human Navigation and the Sixth Sense* (London: Hodder & Stoughton, 1981); R. R. Baker, "Magnetoreception by Humans and Other Primates," in J. L. Kirschvink, D. S. Jones, and B. J. MacFadden (Eds.), *Magnetite Biomineralization and Magnetoreception in Organisms: A New Magnetism* (New York: Plenum, 1985).

23. M. Barinaga, "Giving Personal Magnetism a Whole New Meaning," *Science,* 1992, *256,* 967.

24. Walcott and Lednor, "Bird Navigation" [1]; Emlen, "Migration" [1].

25. Baker, *Bird Navigation* [1]; Gerrard, *Instinctive Navigation* [1].

26. W. T. Keeton, T. S. Larkin, and D. M. Windsor, "Normal Fluctuations in the Earth's Magnetic Field Influence Pigeon Orientation," *Journal of Comparative Physiology,* 1974, *95,* 95–103; C. Walcott, "Anomalies in the Earth's Magnetic Field Increase the Scatter of Pigeons' Vanishing Bearings," in Schmidt-Koenig and Keeton (Eds.), *Animal Migration,* pp. 143–151 [1].

27. W. T. Keeton, "Magnets Interfere with Homing Pigeons," *Proceedings of the National Academy of Sciences,* 1971, *68,* 102–106.

28. J. L. Gould, "The Case for Magnetic Sensitivity in Birds and Bees (Such as It Is)," *American Scientist,* 1980, *68* (No. 3), 256–267; D. Presti and J. D. Pettigrew, "Ferro-magnetic Coupling to Muscle Receptors as a Basis for Geomagnetic Field Sensitivity in Animals," *Nature,* 1980, *285,* 99–101; C. Walcott, J. Gould, and J. Kirschvink, "Pigeons Have Magnets," *Science,* 1979, *205,* 1027–1029.

29. Baker, *The Mystery of Migration* [2]; Baker, *Bird Navigation* [1]; W. Wiltschko and R. Wiltschko, "The Role of Outward Journey Information in the Orientation of Homing Pigeons," in Papi and Wallraff (Eds.), *Avian Navigation,* pp. 239–252 [1].

30. Baker, *Bird Navigation* [1]; W. E. Southern, "Orientation Responses of Ring-billed Gull Chicks: A Re-evaluation," in Schmidt-Koenig and Keeton (Eds.), *Animal Migration,* pp. 311–317 [1]; W. Wiltschko, "The Influence of Magnetic Total Intensity and Inclination on Directions Preferred by Migrating European Robins (*Erithacus rubecula*)," in Galler, Schmidt-Koenig, Jacobs, and Belleville (Eds.), *Animal Orientation and Navigation,* pp. 569–578 [12]; W. Wiltschko, "Der Magnetkompass der Gartengrasmücke (*Sylvia borin*)," *Journal of Ornithology,* 1974, *115,* 1–7; W. Wiltschko,

"The Migratory Orientation of Garden Warblers, *Sylvia borin*," in Papi and Wallraff (Eds.), *Avian Navigation*, pp. 50–58 [1]; W. Wiltschko, and R. Wiltschko, "Magnetic Compass of European Robins," *Science*, 1972, *176*, 62–64.

31. F. Papi, "Olfaction and Homing in Pigeons: Ten Years of Experiments," in Papi and Wallraff (Eds.), *Avian Navigation*, pp. 149–159 [1]. For a rigorous critique of these experiments, which concludes that "some unidentified airborne factor has been shown to have a partial role in homing," see K. Schmidt-Koenig, "Bird Navigation: Has Olfactory Orientation Solved the Problem?" *Quarterly Review of Biology*, 1987, *62*, 31–47.

32. Able, "Mechanisms of Orientation" [1]; T. C. Grubb, "Smell and Foraging in Shearwaters and Petrels," *Nature*, 1972, *237*, 404–405; Keeton, "The Orientation and Navigation of Birds" [1].

33. M. L. Kreithen and D. B. Quine, "Infrasound Detection by the Homing Pigeon," *Journal of Comparative Physiology*, 1979, *129*, 1–4.

34. M. L. Kreithen, "Sensory Mechanisms for Animal Orientation — Can Any New Ones Be Discovered?" in Schmidt-Koenig and Keeton (Eds.), *Animal Migration*, pp. 25–34 [1].

35. Baker, *Bird Navigation* [1]; H. Dingle, "Ecology and Evolution of Migration," in Gauthreaux, Jr. (Ed.), *Animal Migration*, pp. 1–101 [1]; M. M. Nice, "Studies in the Life History of the Song Sparrow," *Transactions of the Linnaean Society of New York*, 1937, *4*, 1–247.

36. Berthold, "Migration" [1]; A. C. Perdeck, "An Experiment on the Ending of Autumn Migration in Starlings," *Ardea*, 1964, *52*, 133–139.

37. Baker, *The Mystery of Migration*, p. 30 [2].

38. V. R. Bingman, "Ontogeny of a Multiple Stimulus Orientation System in the Savannah Sparrow (*Passerculus sandwichensis*)" Ph.D. dissertation, State University of New York at Albany, 1981.

39. Keeton, "The Orientation and Navigation of Birds" [1].

40. Wiltschko and Wiltschko, "The Role of Outward Journey Information" [29].

41. Able, "Mechanisms of Orientation" [1]; A. C. Perdeck, "Orientation of Starlings After Displacement to Spain," *Ardea*, 1967, *55*, 194–202; C. J. Ralph and L. R. Mewaldt, "Homing Success in Wintering Sparrows," *Auk*, 1976, *93*, 1–14.

42. Baker, *Bird Navigation* [1]; Emlen, "Migration" [1]. The Walcott

quotation is from D. Berreby, "Lost Souls," *Discover*, April, 1992, 92–94.

43. Griffin, *Bird Migration* [1].

44. Waterman, *Animal Navigation*, p. 59 [22].

45. D. Lewis, *We, the Navigators* (Canberra: Australian National University Press, 1972); Waterman, *Animal Navigation* [22].

46. I. Eibl-Eibesfeldt, *Human Ethology* (New York: Aldine de Gruyter, 1989). Eibl-Eibesfeldt and other ethologists use the term phylogenetic adaptation instead of instinct even though both mean essentially the same thing. Apparently they wish to drop the term instinct because it has been (wrongly) criticized as connoting rigidity and automaticity. Since instinctive programs are not implemented rigidly or automatically but intelligently (utilizing learning and practice), the term instinct has been dropped for invalid reasons; we should continue to use it and clarify it.

47. Baker, *Bird Navigation*, pp. 202–203 [1].

48. Ibid.; Berthold, "Migration" [1].

49. A. Akmajian, R. A. Demers, and R. M. Harnish, *Linguistics: An Introduction to Language and Communication* (2d Ed.) (Cambridge, Mass.: The MIT Press, 1984).

Chapter 8. Personal Friendships Between Humans and Birds

1. M. S. Corbo and D. M. Barras, *Arnie, the Darling Starling* (Boston: Houghton Mifflin, 1983).

2. W. Steinigeweg, *The New Softbill Handbook* (Hauppauge, N.Y.: Barron's, 1988), p. 88.

3. S. Leek, *The Jackdaw and the Witch* (Englewood Cliffs, N.J.: Prentice-Hall, 1966).

4. B. Heinrich, *One Man's Owl* (Princeton, N.J.: Princeton University Press, 1987).

5. Ibid., p. 16.

6. Ibid., p. 137.

7. Ibid., p. 148.

8. Ibid., p. 55.

9. Ibid., p. 120.

10. Ibid., p. 126.

11. S. C. Wilson, personal communication, September 23, 1990.
12. I. Birmelin and A. Wolter, *The New Parakeet Handbook* (Hauppauge, N.Y.: Barron's, 1985), p. 14.
13. Ibid., p. 86.
14. Ibid., pp. 86–87 [12].
15. J. Kastner, *A World of Watchers: An Informal History of the American Passion for Birds* (San Francisco: Sierra Club, 1986), p. 45.
16. Birmelin and Wolter, *The New Parakeet Handbook* [12].
17. Ibid., p. 33.
18. Ibid., p. 123.
19. L. Howard, *Birds as Individuals* (London: Readers Union, Collins, 1953).
20. Ibid., p. 16.
21. Ibid., p. 24.
22. Ibid., p. 43.
23. Ibid., p. 43–47 [19].
24. Ibid., pp. 130–131.
25. Ibid., pp. 187–188.
26. Ibid., pp. 188–189.
27. Ibid., pp. 18–19.
28. Ibid., pp. 19–20.
29. M. Bishop, *St. Francis of Assisi* (Boston: Little, Brown & Co., 1974), pp. 184–185.
30. E. Underhill, *Mysticism* (New York: Dutton, 1961), pp. 261–262.

Chapter 9. Overview of Bird Intelligence

1. Of course, some kinds of birds live at a much slower tempo than typical avian species. Keep in mind that there are exceptions to virtually every generalization about birds (including their temporal speedup), just as there are exceptions to practically every generalization that can be made about humans.
2. For a succinct introduction to the vast literature of cultural anthropology see L. A. White, "Human Culture," *Encyclopedia Britannica* (15th Ed.), 1977, *8*, 1151–1159. Thorough reviews are presented by A. Kroeber, *Anthropology* (New York: Harcourt Brace, 1948); M. J. Herskovitz, *Man and His Works* (New York: Alfred A. Knopf, 1948); and C. Geertz, *The Interpretation of Cultures* (New York: Basic Books, 1973).

Chapter 10. Why Birds Have Been Totally Misunderstood

1. J. Huxley's Foreword to L. Howard, *Birds as Individuals* (London: Readers Union, Collins, 1953), pp. 9–10.
2. B. L. Hart, *The Behavior of Domesticated Animals* (New York: W. H. Freeman, 1985). Humans have been very effective in changing the basic nature of their enslaved fowl, especially since the 1940s when factory poultry production began. The aim of this factory industry, to produce fast-growing, broad-chested birds, has been fulfilled. Commercial poultry now grows twice as fast as fifty years ago and has a much bigger body. As Jim Mason and Peter Singer point out (*Animal Factories* [New York: Crown, 1990]), the production of cheap meat has made "monsters" of the "butterball" turkeys, which are so "chunky" they cannot carry out sexual intercourse and chickens grow so top-heavy they can hardly walk and die suddenly from the "flipover" syndrome. Since the heavily muscled bodies of the fast-growing chickens need more oxygen than their lungs can provide, the heart enlarges, pressure builds up in the blood vessels to the lungs, fluid leaks into the lungs and body cavity, and they suffer, and some die, from ascites, or "waterbelly." Since the weight of their fast-growing bodies overwhelms their scrawny legs, they spend much of their time crouching, which causes sores that easily become infected and kill the birds.
3. L. J. Stettner and K. A. Matyniak, "The Brain of Birds," in B. B. Wilson (Ed.), *Birds: Readings from Scientific American* (San Francisco: W. H. Freeman, 1980), pp. 192–199.
4. Hart, *The Behavior of Domesticated Animals* [2]; A. M. Guhl, "The Social Order of Chickens," *Scientific American*, February 1956, *194* (No. 2), 42–46; T. Schjelderup-Ebbe, "Social Behavior in Birds," in C. Murchison (Ed.), *Handbook of Social Psychology* (Worcester, Mass.: Clark University Press, 1935).
5. G. S. Carter, *A General Zoology of the Invertebrates* (London: Sidgwick-Jackson, 1948); E. M. Macphail, *Brain and Intelligence in Vertebrates* (New York: Oxford University Press, 1982); E. Step, *Marvels of Insect Life,* (New York: National Travel Club, 1938).
6. Macphail, *Brain and Intelligence in Vertebrates* [5].
7. F. Nottebohm, "From Bird Song to Neurogenesis," *Scientific American*, February 1989, *260* (No. 2), 74–79.

8. H. Gardner, *Frames of Mind: The Theory of Multiple Intelligences* (New York: Basic Books, 1983).

9. C. Darwin, *The Descent of Man and Selection in Relation to Sex* (London: John Murray, 1871); C. Darwin, *The Expression of Emotions in Man and Animals* (London: John Murray, 1872).

10. G. J. Romanes, *Animal Intelligence* (New York: D. Appleton & Co., 1883).

11. Ibid., pp. 270–271.

12. Ibid., pp. 274–275.

13. Ibid., p. 276.

14. Ibid., pp. 318–319.

15. Ibid., p. 319.

16. E. G. Boring, *A History of Experimental Psychology* (New York: Appleton-Century-Crofts, 1950); B. E. Rollin, *The Unheeded Cry: Animal Consciousness, Animal Pain and Science* (New York: Oxford University Press, 1989); S. Walker, *Animal Thought* (London: Routledge & Kegan Paul, 1983).

17. T. X. Barber, *Pitfalls in Human Research: Ten Pivotal Points* (Elmsford, N.Y.: Pergamon, 1976); R. Boakes, *From Darwin to Behaviourism: Psychology and the Minds of Animals* (New York: Cambridge University Press, 1984); Boring, *A History of Experimental Psychology* [16]; Walker, *Animal Thought* [16].

Chapter 11. Are All Animals Intelligent?

1. American Sign Language (and other comparable languages of the deaf) is linguistically equivalent to the spoken languages used by hearing individuals. An excellent recent discussion of this equivalence is found in O. Sacks, *Seeing Voices: A Journey into the World of the Deaf* (Berkeley: University of California Press, 1989).

2. The material on Koko is from: F. G. Patterson, *Linguistic Capabilities of a Lowland Gorilla* (Ann Arbor, Mich.: University Microfilms, 1979) [Ph.D. dissertation, Stanford University, No. 79-17269]; F. G. Patterson and E. Linden, *The Education of Koko* (New York: Holt, Rinehard, & Winston, 1981); and Vols. 8, 11, and 12 (1984–85, 1987–88, 1988–89) of *Gorilla: Journal of the Gorilla Foundation* (Woodside, Calif.).

3. Innovative techniques of teaching an ape sign language, for instance, by "molding" the hands, were pioneered by Alan and

Beatrix Gardner whose work with the chimpanzee Washoe, discussed in the next section of this chapter, preceded and stimulated Patterson's work with Koko.

4. The totally unconvincing attempt to explain the performance of Koko, Washoe, and other sign-using apes as due to unconscious bias on the part of the experimenters is found in J. Umiker-Sebeok and T. A. Sebeok, "Clever Hans and Smart Simians," *Anthropos*, 1981, *76*, 89–165. The Sebeoks failed to convince the serious students of chimpanzee-human communication that it was due to an unnoticed and unwanted experimenter effect.

5. The material on Washoe derives from a recent publication—R. A. Gardner, B. T. Gardner, and T. E. Van Cantfort (Eds.), *Teaching Sign Language to Chimpanzees* (Albany: State University of New York Press, 1989)—and from earlier supplemental articles: B. T. Gardner and R. A. Gardner, "Two-way Communication with an Infant Chimpanzee," in A. M. Schrier and F. Stollnitz (Eds.), *Behavior of Nonhuman Primates*, Vol. 4 (New York: Academic Press, 1971), pp. 117–184; B. T. Gardner and R. A. Gardner, "Comparing the Early Utterances of Child and Chimpanzee," in A. Pick (Ed.), *Minnesota Symposium on Child Psychology*, Vol. 8 (Minneapolis: University of Minnesota Press, 1974), pp. 3–23; B. T. Gardner and R. A. Gardner, "Evidence for Sentence Constituents in the Early Utterances of Child and Chimpanzee," *Journal of Experimental Psychology: General*, 1975, *104*, 244–267; R. A. Gardner and B. T. Gardner, "Comparative Psychology of Language Acquisition," *Annals of the New York Academy of Sciences*, 1978, *309*, 37–76.

6. E. S. Savage-Rumbaugh, *Ape Language: From Conditioned Response to Symbol* (New York: Columbia University Press, 1986).

7. S. Savage-Rumbaugh, M. A. Romski, W. D. Hopkins, and R. A. Sevcik, "Symbol Acquisition and Use by *Pan troglodytes, Pan paniscus, Homo sapiens*," in P. G. Heltine and L. A. Marquardt (Eds.), *Understanding Chimpanzees* (Cambridge, Mass.: Harvard University Press, 1989), pp. 266–295.

8. S. Savage-Rumbaugh, "Communication, Symbolic Communication, and Language: Reply to Seidenberg and Petitto," *Journal of Experimental Psychology: General*, 1987, *116*, 288–292.

9. F. de Waal, *Chimpanzee Politics: Power and Sex Among Apes* (New York: Harper, 1982). The quotation is from the Foreword by Desmond Morris, pp. 13–15.

10. J. Goodall, *The Chimpanzees of Gombe: Patterns of Behavior* (Cam-

bridge, Mass.: Harvard University Press, 1986). The quotation describing the "special parties" is from page 151.

Many other careful observers have also emphasized the humanlike behavior of chimpanzees; see, for instance, D. O. Hebb, "Emotion in Man and Animal: An Analysis of the Intuitive Processes of Recognition," *Psychological Review*, 1946, *53*, 88–106. In a later review of animal studies, Hebb commented on his earlier paper as follows:

> It has been noted elsewhere . . . that exposure to a group of adult chimpanzees gives one the overwhelming conviction that one is dealing with an essentially human set of attitudes and motivations. The feeling is not less strong in the psychologically trained than in the untrained (and its strength can be embarrassing to the purist who thinks that naming an animal's attitude or emotion is "subjective" and commits the deadly sin of anthropomorphism). It is probably a common experience to all who have worked at the Yerkes Laboratories to feel that the bare bones of human personality, the raw essentials are being laid open before his eyes.

D. O. Hebb and W. R. Thompson, "The Social Significance of Animal Studies," in G. Lindzey and E. Aronson (Eds.), *The Handbook of Social Psychology, Vol. 2 Research Methods* (Reading, Mass.: Addison-Wesley, 1968), pp. 729–774. The quotation is from page 747.

11. L. M. Herman, D. G. Richards, and J. P. Wolz, "Comprehension of Sentences by the Bottlenosed Dolphin," *Cognition*, 1984, *16*, 129–219; L. M. Herman, "Cognition and Language Competencies of Bottlenosed Dolphins," in R. G. Schusterman, J. Thomas, and F. G. Wood (Eds.), *Dolphin Cognition and Behavior: A Comparative Approach* (Hillsdale, N.J.: Lawrence Erlbaum Associates, 1986), pp. 221–252; L. M. Herman, "Receptive Competencies of Language-trained Animals," in J. S. Rosenblatt (Ed.), *Advances in the Study of Behavior*, Vol. 17 (New York: Academic Press, 1987), pp. 1–60.

12. L. M. Herman, "What the Dolphin Knows, or Might Know, in Its Natural World," in K. Pryor and K. S. Norris (Eds.), *Dolphin*

Societies: Discoveries and Puzzles (Berkeley: University of California Press, 1991), pp. 349–363.

13. K. W. Pryor, R. Haag, and J. O'Reilly, "The Creative Porpoise: Training for Novel Behavior," *Journal of the Experimental Analysis of Behavior*, 1969, *12*, 653–661. L. M. Herman reports that his research team has verified this creative capability with bottlenosed dolphins (Herman, "What the Dolphin Knows" [12]). The dolphins were stimulated to behave creatively in essentially the same way they learned symbols for words, by receiving food rewards when they did so.

14. The quotation is from W. Booth, "The Joys of a Big Brain: Cooperation, Friendship, Sex and Pleasure Are Very Human and Very Dolphin Too," *Psychology Today*, April 1989, p. 57. The original material is presented in the following: R. S. Wells, "The Role of Long-term Study in Understanding the Social Structure of a Bottlenose Dolphin Community," in Pryor and Norris (Eds.), *Dolphin Societies*, pp. 199–225 [12]; R. C. Connor and K. S. Norris, "Are Dolphins Reciprocal Altruists?" *American Naturalist*, 1982, *119* (No. 3), 358–374; C. M. Johnson and K. S. Norris, "Delphinid Social Organization and Social Behavior," in Schusterman, Thomas, and Wood (Eds.), *Dolphin Cognition and Behavior*, pp. 335–346 [11]; B. Würsig, "Delphinid Foraging Strategies," in Schusterman, Thomas, and Wood (Eds.), *Dolphin Cognition and Behavior*, pp. 347–359 [11]; J. W. Bradbury, "Social Complexity and Cooperative Behavior in Delphinids," in Schusterman, Thomas, and Wood (Eds.), *Dolphin Cognition and Behavior*, pp. 361–372 [11]; K. Pryor and I. K. Shallenberger, "Social Structure in Spotted Dolphins (*Stenella attenuata*) in the Tuna Purse Seine Fishery in the Eastern Tropical Pacific," in Pryor and Norris, *Dolphin Societies*, pp. 161–196 [12].

15. Herman, "What the Dolphin Knows" [12].

16. P. Froiland, "Understanding Dolphins," *Passages*, December 1982, 45–50. The quotation is from page 48.

17. The first paper to delineate the musical characteristics of whale songs is R. S. Payne and S. McVay, "Songs of Humpback Whales," *Science*, 1971, *173*, 585–597. Another important early paper, reporting that a whale may continue singing for at least twenty-two hours, is by H. E. Winn and I. K. Winn, "The Song of the Humpback Whale (*Megaptera novaeangliae*) in the West Indies,"

Marine Biology, 1978, *47*, 97–114. Subsequent work on whale musical intelligence is presented in the following papers that are included in R. Payne (Ed.), *Communication and Behavior of Whales* (Boulder, Colo.: Westview Press, 1983): K. Payne, P. Tyack, and R. Payne, "Progressive Changes in the Songs of Humpback Whales (*Megaptera novaeangliae*): A Detailed Analysis of Two Seasons in Hawaii," pp. 9–57; L. N. Guinee, K. Chu, and E. M. Dorsey, "Changes Over Time in the Songs of Known Individual Humpback Whales (*Megaptera novaeangliae*)," pp. 59–80; P. Frumhoff, "Aberrant Songs of Humpback Whales (*Megaptera novaeangliae*): Clues to the Structure of Humpback Songs," pp. 81–127; R. Payne and L. Guinee, "Humpback Whale (*Megaptera novaeangliae*) Songs as an Indicator of Stocks," pp. 333–358.

18. Roger Payne's statement is quoted in R. F. Burgess, *Secret Languages of the Sea* (New York: Dodd, Mead & Co., 1981), pp. 157–158.

19. The rhymelike pattern was discovered by Linda Guinee and Katherine Payne (see *Discover*, July 1989, p. 22).

20. G. Fryer and T. D. Iles, *The Cichlid Fishes of the Great Lakes of Africa* (Edinburgh: Oliver & Boyd, 1972).

21. Ibid.

22. W. B. Vernberg, *Symbiosis in the Sea* (Columbia: University of South Carolina Press, 1973); G. S. Losey, Jr., "The Symbiotic Behavior of Fishes," in D. J. Mostofsky (Ed.), *Behavior of Fish and Other Aquatic Animals* (New York: Academic Press, 1978).

23. C. Limbaugh, "Cleaning Symbiosis," *Scientific American*, August 1961, 42–49.

24. Ibid.

25. L. R. Aronson, "Orientation and Jumping Behavior in the Gobiid Fish, *Bathygobius soporator*," *American Museum Novitiate*, 1951, *1486*, 1–22; L. R. Aronson, "Further Studies in Orientation and Jumping Behavior in the Goby Fish, *Bathygobius soporator*," *Anatomical Record*, 1956, *125*, 606; L. R. Aronson, "Further Studies on Orientation and Jumping Behavior in the Gobiid Fish, *Bathygobius soporator*," *Annals of the New York Academy of Sciences*, 1971, *188*, 378–392.

26. The earlier literature is summarized in G. J. Romanes, *Animal Intelligence* (New York: Appleton, 1883), pp. 241–253.

27. W. M. Wheeler, *Ants: Their Structure, Development, and Behavior* (New York: Columbia University Press, 1910), p. 535.

28. B. Hölldobler and E. O. Wilson, *The Ants* (Cambridge, Mass.: Harvard University Press, 1990), p. 227.

29. The translations pertain to the language of the African weaver ant (*Oecophylla longinoda*) as described in ibid., p. 251.

30. The symbolic nature of this "chain communication" was pointed out by D. R. Griffin, *Animal Thinking* (Cambridge, Mass.: Harvard University Press, 1984), p. 172.

31. See the discussion on "pheromone blends" in Hölldobler and Wilson, *The Ants*, pp. 246–249 [28].

32. Examples of different responses to the same messages are common in the myrmecological literature; see, for example, Hölldobler and Wilson, *The Ants*, pp. 255 and 339 [28].

33. Hölldobler and Wilson, *The Ants*, pp. 277 and 296 [28].

34. *Pogonomyrmex* species. B. Hölldobler, "Ethological Aspects of Chemical Communication in Ants," *Advances in the Study of Behavior*, 1978, *8*, 75–115, p. 81.

35. W. S. Creighton and M. P. Creighton, "The Habits of *Pheidole militicida* Wheeler (Hymenoptera: Formicidae)," *Psyche*, 1959, *66* (No. 1–2), 1–12.

36. E. O. Wilson, *Sociobiology: The New Synthesis* (Cambridge, Mass.: Harvard University Press, 1975), p. 549.

37. R. Chauvin, *The World of Ants* (New York: Hill & Wang, 1971), p. 122.

38. Wilson, *Sociobiology*, p. 549 [36].

39. H. C. Van der Heyde, "Quelques observations sur la psychologie des fourmis," *Archives Néerlandais physiologie*, 1920, *4*, 259–264.

40. Hölldobler and Wilson, *The Ants*, p. 342 [28].

41. There are many reports of army ants using their bodies to bridge extensive spans and a smaller number of reports of ants building tunnels to reach a goal: Chauvin, *The World of Ants*, p. 206 [37]. Various kinds of ants have also been observed to build a bridge bit by bit across a liquid barrier. Romanes, *Animal Intelligence*, pp. 135–137 [26]; W. R. Corliss, *Incredible Life: A Handbook of Biological Mysteries* (Glen Arm, Md.: The Sourcebook Project, 1981), pp. 692–693.

Building a bridge step by step to reach a goal requires not only intelligent awareness but also deliberate, purposeful actions of the kind that have been considered uniquely human. Consequently, to comply with the official science dogma, which forbids attribution of humanlike characteristics to ants, myrmecologists must brush aside any evidence that ants can construct a bridge in a goal-directed way one piece at a time. The dean of present-day myrmecologists, E. O. Wilson, for instance, brusquely dismisses all earlier reports: "Ant workers also occasionally try to cover small pools of water or other liquid in the nest vicinity. The casual observation of this phenomenon has misled some authors to report erroneously that ants construct 'bridges' to cross obstacles" (*The Insect Societies* [Cambridge, Mass.: Harvard University Press, 1971], p. 278). (These exact words are also repeated in Hölldobler and Wilson, *The Ants*, p. 296 [28].) However, Wilson is mistaken in dismissing reports of ant bridges as due to authors that were "misled" by "causal observations." The reports were vouched for by distinguished European scientists who were not easily misled. The ants that were observed did not simply try to cover pools of water; instead, they placed appropriate material bit by bit where it would bridge various kinds of barriers. The reports summarized in the publications of the entomologist R. A. Réaumur, the zoologist Büchner, the zoologist Karl Vogt, and the academician Leuckart state that ants observed in various parts of Europe constructed bridges across barriers by the following methods: placing many tiny pieces of wood in line (as if they had "knowledge of practical engineering"); lining up a tiny splinter; placing aphids "on the tar one after another until they made a bridge"; pulling and pushing a straw until it formed a bridge where it was needed; and procuring "little pellets of earth, which they carried in their jaws and deposited one after another upon the [liquid barrier] till a road of earth was made across it." (Romanes, *Animal Intelligence*, pp. 135–137 [26].) The conclusion indicated is that ants, like humans, can intelligently and purposively construct bridges to reach their goals.

42. The startling ability of ants to learn a maze with ten blind alleys was demonstrated by T. C. Schneirla in a series of experiments discussed in M. R. Washburn, *The Animal Mind: A Text-Book of Comparative Psychology* (New York: Macmillan, 1936), p. 295. The ants' ability to remember their way through mazes when tested four days later was demonstrated by R. Chauvin, "Expériences sur l' 'apprentissage par équipe' due labyrinthe chez *Formica polyctena*," *Insectes Sociaux*, 1964, *11* (No. 1), 1–20.

43. Hölldobler and Wilson, *The Ants*, p. 367 [28]; R. Wehner and R.

Menzel, "Do Insects Have Cognitive Maps?" *Annual Review of Neuroscience*, 1990, *13*, 403–414.

44. The species is *Solenopsis molesta*. Wheeler, *Ants*, p. 427 [27]; Wilson, *Insect Societies*, p. 357 [41].

45. Wheeler, *Ants*, pp. 192–224. [27]. An example is the "crazy ant," *Anoplolepis longipes*, which may build its nest in the soil, under rocks, under fallen logs, under leaves, in the crowns of coconut palms— virtually anywhere. J. H. Haines and J. B. Haines, "Colony Structure, Seasonality, and Food Requirements of the Crazy Ant, *Anoplolepis longipes*, in the Seychelles," *Ecological Entomology*, 1978, *3*, 109–118.

46. Wheeler, *Ants*, p. 195 [27].

47. Hölldobler and Wilson, *The Ants*, pp. 171–174 [28].

48. Wheeler, *Ants*, pp. 195–196, 222–223 [27].

49. H. C. McCook, *Ant Communities* (New York: Harper, 1909), pp. 40, 50.

50. The summary of "every kind of warfare known to ourselves" that is also found in ants is from M. Maeterlinck, *The Life of the Ant* (New York: John Day Co., 1930), pp. 107–108.

51. The early report of an ant tournament is by the "father of myrmecology," P. Huber, as quoted by Maeterlinck in *The Life of the Ant* (p. 91) [50]. The very recent report by the two foremost present-day myrmecologists is by Hölldobler and Wilson, *The Ants*, pp. 410–411 [28]. A complementary recent report is B. Hölldobler, "Tournaments and Slavery in a Desert Ant," *Science*, 1976, *192*, 912–914.

52. Hölldobler and Wilson, *The Ants*, pp. 596–608 [28].

53. Wheeler, *Ants*, pp. 223–224 and 339–360 [27]; M. V. Brian, *Social Insects: Ecology and Behavioural Biology* (London: Chapman & Hall, 1983), p. 22.

54. Wheeler, *Ants*, pp. 452–486 [27].

55. Forel's observations are quoted in detail by Romanes, *Animal Intelligence*, pp. 70–76 [26].

56. Ibid., p. 122 [26].

57. Hölldobler and Wilson, *The Ants*, pp. 631–633 [28].

58. Wheeler, *Ants*, pp. 509 and 529 [27].

59. The following publications were especially helpful in integrating the rich data now available pertaining to the intelligence of honeybees: K. von Frisch, *The Dance Language and Orientation of Bees*

(Cambridge, Mass.: Harvard University Press, 1967); J. L. Gould and C. G. Gould, *The Honey Bee* (New York: Scientific American Library, 1988); M. Lindauer, *Communication Among Social Bees* (Cambridge, Mass.: Harvard University Press, 1961); T. D. Seeley, *Honeybee Ecology* (Princeton, N. J.: Princeton University Press, 1985). Supplementary material was obtained from: C. D. Michener, *The Social Behavior of the Bees* (Cambridge, Mass.: Harvard University Press, 1974); M. Winston, *The Biology of the Honey Bee* (Cambridge, Mass.: Harvard University Press, 1987).

60. T. D. Seeley, "How Honeybees Find a Home," *Scientific American,* October 1982, *247* (No. 4), 158–168. Also summarized in Seeley, *Honeybee Ecology,* pp. 71–75 [58].

61. V. Frisch, *The Dance Language* [58]; Lindauer, *Communication Among Social Bees* [58]; Gould and Gould, *The Honey Bee* [58]; Seeley, *Honeybee Ecology* [58].

62. H. Markl, "Manipulation, Modulation, Information, Cognition: Some of the Riddles of Communication," in B. Hölldobler and M. Lindauer (Eds.), *Experimental Behavioral Ecology and Sociobiology* (Sunderland, Mass.: Sinauer Associates, 1985), pp. 163–194. The quotation is from p. 189.

63. The bees in the hive perceive the scout's dance movements, not visually, but by touching the dancer and hearing the vibratory waggles. Parenthetically, if the new source of food, water, or resin is nearby (less than roughly fifty meters from the hive), the scout performs a "round dance," which is essentially the same as the "waggle dance" without the waggles.

64. Seeley, *Honeybee Ecology,* pp. 83–84 [58].

65. Gould and Gould, *The Honey Bee,* pp. 59–61 [58]. D. R. Griffin's comments place the discovery of the honeybee language in a broader historical context:

> In the scientific climate of opinion prevailing forty years ago it was shocking and incredible to be told that a mere insect could communicate to its companions the direction, distance and desirability of something far away. At least in America, we were still firmly under the spell of behaviorism in psychology and of the reductionistic approach to animal behavior. . . . But when we learned from the classic experiments of Lindauer (1955) that semantically communicative ex-

changes are used by swarming bees to reach a consensus, the superiority of not only our species but our phylum seemed to be challenged. . . . von Frisch's Rosetta Stone has certainly narrowed the gap between human language and animal communication; and of course human language has always been considered to be closely linked to human thinking.

D. R. Griffin, "The Cognitive Dimensions of Animal Communication," in Hölldobler and Lindauer (Eds.), *Experimental Behavioral Ecology and Sociobiology*, pp. 471–473 [62].
66. H. Esch, "The Evolution of Bee Language," *Scientific American*, 1967, *216* (No. 4), 97–104.
67. M. Maeterlinck, *The Life of the Bee* (New York: Dodd, Mead, & Co., 1901).
68. Romanes, *Animal Intelligence*, p. 207 [26].
69. C. Darwin, *The Origin of Species by Means of Natural Selection* (New York: Random House, 1950 [1859]), p. 187; Maeterlinck, *The Life of the Bee*, pp. 363–368 [66].
70. Gould and Gould, *The Honey Bee*, pp. 221–222 [58].
71. Ibid., pp. 125–155; Wehner and Menzel, "Do Insects Have Cognitive Maps?" [43].

Chapter 12. Revolutionary Implications of Animal Intelligence

1. P. Sherrard, *The Eclipse of Man and Nature: An Enquiry into the Origins and Consequences of Modern Science* (West Stockbridge, Mass.: Lindisfarne Press, 1987), p. 93.
2. An example of natural ecological recycling is seen in Africa's Serengeti. P. R. Ehrlich, *The Machinery of Nature* (New York: Simon & Schuster, 1986), pp. 239–260. The plants are consumed by the herbivores in the most proficient way. The zebras come first; then the wildebeest arrive and feed on leaves made more accessible by the zebras. Then Thomson's gazelles feed on herbs made more accessible by the earlier herbivores. Now enter the carnivores— lions, leopards, cheetahs, hyenas, and others—which divide the herbivores among themselves with an amazing proficiency by

hunting at different times (day or night) and by preferring prey of different sizes. Finally enter the decomposers—vultures, insects, fungi, bacteria—each with its different specialty in breaking down different components of the carrion. The chemicals from the decomposition are needed as nutrients by the plants, thus bringing the food chain around full circle.

3. T. Kuhn, *The Structure of Scientific Revolution* (Chicago: University of Chicago Press, 1962). Scientists who uncritically accept the dominant paradigm that views animals as much more like machines than like people are the focus of discussion in B. E. Rollin, *The Unheeded Cry: Animal Consciousness, Animal Pain, and Science* (New York: Oxford University Press, 1989). See also M. J. Mahoney, *Scientist as Subject: The Psychological Imperative* (Cambridge, Mass.: Ballinger Publishing Co., 1976).

4. L. Howard, *Birds as Individuals* (London: Readers Union, Collins, 1953), p. 13.

5. M. Fox, *The Coming of the Cosmic Christ: The Healing of Mother Earth and the Birth of a Global Renaissance* (San Francisco: Harper & Row, 1988), p. 14; J. Terborgh, *Where Have all the Birds Gone?* (Princeton, N.J.: Princeton University Press, 1990); J. Bohlen, "New Threat of a Silent Spring," *Defenders*, November–December 1984, *59*, 20–29; W. B. King (Ed.), *Endangered Birds of the World* (Washington, D.C.: Smithsonian Institution, 1981).

6. The data in this section are based primarily on the following: Many useful reports on the state of the world by researchers at the Worldwatch Institute in Washington, D.C. (including L. R. Brown, C. Flavin, S. Postel, W. Renner, J. L. Jacobson, A. B. Durning, C. P. Shea, and E. C. Wolf); T. Berry, *The Dream of the Earth* (San Francisco: Sierra Club Books, 1988); A. Gordon and D. Suzuki, *It's a Matter of Survival* (Cambridge, Mass.: Harvard University Press, 1991); Fox, *The Coming of the Cosmic Christ* [5]; B. Tokar's many thoughtful articles on environmental problems in *Z Magazine* and his overview in *The Green Alternative: Creating an Ecological Future* (San Pedro, Calif.: R. & E. Miles, 1987); B. Devall and G. Sessions, *Deep Ecology: Living as if Nature Mattered* (Layton, Utah: Gibbs M. Smith, 1985); B. McKibben, *The End of Nature* (New York: Random House, 1989); P. Ehrlich and A. Ehrlich, *Extinction: The Causes and Consequences of the Disappearance of Species* (New York: Ballantine, 1981); J. Rifkin, *Biosphere Politics* (New

York: Crown, 1991); many articles on the environment by K. Sale in political journals such as *The Nation* (April 30, 1990; May 14, 1988); many useful articles in such journals as *Whole Earth, Greenpeace, Utne Reader, E Magazine,* and *Z Magazine.* Two critiques of the data in this area were also useful: D. L. Ray and L. Guzzo, *Trashing the Planet* (Washington, D.C.: Regnery Gateway, 1990); J. L. Simon and H. Kahn (Eds.), *Resourceful Earth* (Oxford, England: Basil Blackwell, 1984).

7. Tokar, *The Green Alternative* [6]. The data on the contamination of U.S. aquifers, the Rhine River, and the Mediterranean Sea are presented by Fox, *The Coming of the Cosmic Christ,* pp. 13–17 [5].

8. R. Gelbspan, "Environmental Notebook," *The Boston Globe,* April 23, 1991, p. 3. This article reports data presented by the National Environmental Law Center and the U.S. Public Interest Research Group.

9. Gordon and Suzuki, *It's a Matter of Survival* [6].

10. A typical report is presented in *Science News,* April 6, 1991, *139,* 212.

11. D. Meadows, "Consequences Greater than the Gulf War," *Annals of Earth,* 1991, *9* (No. 1), 27; McKibben, *The End of Nature,* p. 23 [6]; J. Leggett (Ed.), *Global Warming: The Greenpeace Report* (New York: Oxford University Press, 1990).

12. G. Medvedev, *The Truth About Chernobyl* (New York: Basic Books, 1991); H. Wasserman, "Chernobyl's American Fallout," *Z Magazine,* June 1989, 60–68.

13. D. De Sante and G. Geupel reported in *The Condor* that there was a dramatic reduction in the number of young birds in the area north of San Francisco during that time and the avian deaths appeared to "coincide remarkably well with the passage of a radioactive 'cloud' from the Chernobyl nuclear power plant accident and associated rainfall." A summary of the De Sante and Geupel report is presented in Wasserman, "Chernobyl's American Fallout" [12].

14. J. Schor, "Capitalism: Triumphant or Putrifying?" *Z Magazine,* March 1990, 52–54.

15. The data on the 50,000 toxic dump sites, the Love Canal study, the doubling of birth defects, and the depletion of the rich Iowa soil and the earth's fertile soil are from Fox, *The Coming of the Cosmic Christ,* pp. 13–17 [5].

16. Berry, *The Dream of the Earth* [6].

17. Fox, *The Coming of the Cosmic Christ*, p. 34 [5].
18. Ehrlich, *The Machinery of Nature*, pp. 17–18 [2].
19. James Lovelock's statement is presented in A. Vittachi, *Earth Conference One: Sharing a Vision for Our Planet* (Boston: Shambhala, 1989), p. 14.
20. E. Sahtouris, *Gaia: The Human Journey from Chaos to Cosmos* (New York: Simon & Schuster, 1989), pp. 228–229. Many other scholars have concluded that a basic change in mankind's attitude to nature is needed to solve our environmental problems. See, for example, the numerous quotations in Devall and Sessions, *Deep Ecology* [6], Gordon and Suzuki, *It's a Matter of Survival* [6], and R. Sheldrake, *The Rebirth of Nature: The Greening of Science and God* (New York: Bantam, 1991).

Appendix A. The Continuing Cognitive Ethology Revolution: A Clarification for My Colleagues in the Behavioral and Brain Sciences

1. D. R. Griffin, *The Question of Animal Awareness: Evolutionary Continuity of Mental Experience* (New York: Rockefeller University Press, 1976); D. R. Griffin, "Prospects for a Cognitive Ethology," *Behavioral and Brain Sciences*, 1978, *1*, 527–538, 555, 609–614, and 1980, *3*, 618–620; D. R. Griffin, "Introduction," in D. R. Griffin (Ed.), *Animal Mind–Human Mind* (New York: Springer-Verlag, 1982), pp. 1–12; D. R. Griffin, "Prospects for a Cognitive Ethology," in J. de Luce and H. T. Wilder (Eds.), *Language in Primates: Perspectives and Implications* (New York: Springer-Verlag, 1983), pp. 159–186; D. R. Griffin, "Thinking About Animal Thoughts," *Behavioral and Brain Sciences*, 1983, *3*, 364; D. R. Griffin, *Animal Thinking* (Cambridge, Mass.: Harvard University Press, 1984); D. R. Griffin, "Epilogue: The Cognitive Dimensions of Animal Communication," in B. Hölldobler and M. Lindauer (Eds.), *Experimental Behavioral Ecology and Sociobiology* (Sunderland, Mass.: Sinauer Associates, 1985), pp. 470–482; D. R. Griffin, "Phylogenetically Widespread 'Facts of Life,'" *Behavioral and Brain Sciences*, 1987, *10*, 667–668; D. R. Griffin, "Subjective Reality," *Behavioral and Brain Sciences*, 1988, *11*, 256; D. R. Griffin, "Progress Towards a Cogni-

tive Ethology," in C. A. Ristau (Ed.), *Cognitive Ethology: The Minds of Other Animals* (Hillsdale, N. J.: Lawrence Erlbaum Associates, 1991), pp. 3–17.

2. The conclusion that animals think, at least in a simple cause-effect way, was predictable from the analysis of causality by Hume (and its amplification by Kant). This analysis led to four conclusions that have never been seriously challenged:

   a. Our notion of cause-effect is a priori [innate or instinctual].

   b. Our ability to infer causes for events is absolutely necessary for our survival. (Without the ability to infer causes for the events we perceive and to infer that we can do something that will cause something else to happen, our world will be in total disarray; there would be no meaning in anything and we would understand nothing.)

   c. The process of inferring cause-effect relations (if this, then that) is indistinguishable from simple thinking.

   d. Animals must also have this ability to infer simple cause-effect relations, to think at least in a simple way, because they too need it to survive.

   To this I add that, since animals possess a basic, essential part of human mentality—the ability to make cause-effect inferences and thus to think at least in a simple way—they are, to this extent, like humans; therefore, the anthropomorphic contention, that animals possess characteristics that humans consider unique to humans, is essentially valid.

3. T. S. Kuhn, *The Structure of Scientific Revolutions* (Chicago: University of Chicago Press, 1962); I. Lakatos, *The Methodology of Scientific Research Programmes* (New York: Cambridge University Press, 1978); I. Lakatos and A. Musgrave (Eds.), *Criticism and the Growth of Knowledge* (New York: Cambridge University Press, 1974); L. Laudan, *Science and Hypothesis* (Dordrecht: D. Reidel, 1981); L. Laudan, *Science and Values: The Aims of Science and their Role in Scientific Debate* (Berkeley: University of California Press, 1984).

   The antianthropomorphic dogma of official science, like the grin of the Cheshire cat, continues to exist even though its supposed basis has disappeared long ago. Sophisticated scientists find the antianthropomorphic dogma ridiculous. For instance, one of the greatest comparative psychologists of the past generation, Margaret F. Washburn, wrote:

All psychic interpretation of animal behavior must be on the analogy of human experience. We do not know the meaning of such terms as perception, pleasure, fear, anger, visual sensation, etc., except as these processes form a part of the contents of our own minds. Whether we will or no, we must be anthropomorphic in the notions we form of what takes place in the mind of an animal.

The Animal Mind (2d Ed.) (New York: Macmillan, 1917), p. 13. After presenting this quotation from Washburn, two sophisticated philosophers, whose specialty is the analysis of research on animal mentality, commented:

It is widely believed that anthropomorphism in general is a fallacy.... Yet when it comes to stating precisely in what the fallacy consists, or explaining why it is a fallacy, the antianthropomorphic forces are strangely silent. As the geneticist A. D. Darbishire noted, the word originally meant "the endowing of God with the form and habits of man.... But those who were responsible ... for applying the word anthropomorphic to an entirely different thing— the granting of intelligence, purpose, design, and human attributes in general to non-human animals ... were the unconscious perpetrators of a successful fraud."... The fraud lies not in extending the term to animals, but in retaining the negative connotations without giving any justification for doing so. Given the framework of evolutionary biology, it is difficult to see what justification there can be for a general indictment of anthropomorphism.

D. Radner and M. Radner, *Animal Consciousness* (Buffalo, N.Y.: Prometheus Books, 1989), p. 140.

4. Many scientists who are not strongly committed to the dominant paradigm will nevertheless continue to adhere to it because of "professional inertia." This predictable effect has been described by geoscientist Edward Bullard, who was a major participant in another scientific revolution (the plate tectonic revolution in the earth sciences):

There is always a strong inclination for a body of professionals to oppose an unorthodox view. Such a group has a considerable investment in orthodoxy; they have learned to interpret a large body of data in terms of the old view, and they have prepared lectures and perhaps written books with the old background. To think the whole subject through again when one is no longer young is not easy and involves admitting a partially misspent youth. . . . Clearly it is more prudent to keep quiet, to be a moderate defender of orthodoxy, or to maintain that all is doubtful, sit on the fence, and wait in statesmanlike ambiguity for more data.

"The Emergence of Plate Tectonics: A Personal View," *Annual Review of Earth and Planetary Science*, 1975, 3, 1–30. The quotation is from p. 5.

Appendix B. How You Can Personally Experience a Bird as an Intelligent Individual

1. L. Howard, *Birds as Individuals* (London: Readers Union, Collins, 1953), pp. 13–15.
2. Ibid., p. 15.
3. Ibid., p. 16.
4. Ibid., p. 18.
5. Ibid., pp. 160–161.
6. Ibid., pp. 166–167.
7. Ibid., pp. 169–170.
8. Ibid., pp. 183, 194.
9. To avoid waste and minimize cleanup of leftover shells, use such food items as cracked corn, unsalted peanut butter, thistle seeds, peanut hearts, and sunflower seeds without shells. See the pamphlets of the Audubon Workshop (Northbrook, Ill.), and R. B. Burton, *National Audubon Society North American Birdfeeder Handbook: The Complete Guide to Attracting, Feeding, and Observing Birds in Your Yard* (New York: Dorling Kindersley, 1992), and John V. Dennis, *A Complete Guide to Bird Feeding* (New York: Knopf, 1986).
10. Personal communication from S. C. Wilson, September 18, 1991.

11. The stress experienced by a home-bred bird when it is moved from its original home to a new home is immensely magnified when a wild bird living freely in a distant place, such as South America or Africa, is kidnapped, imprisoned, shipped in a crammed crate over long distances, and quarantined. Since it is unlikely that such a traumatized bird will lose its fear and mistrust of people, it is unlikely that the purchaser of such a "pet" will perceive its intelligence. What a pity that so many birds sold in pet stores are imported and traumatized. As a buyer, you should ascertain, when purchasing a bird, that it has not been imported and that it has had positive associations with people. (Also, try to obtain a bird that has been hand fed and visit it to bond with it before you take it home.)

12. After Blue Bird felt secure in his new home, he was introduced to strange sounds and other potentially frightening stimuli gradually and carefully while Wilson simultaneously spoke comforting words of reassurance.

13. Parakeets in nature (living in the dry Australian interior) do not have permanent homes; instead, they continually roam with their flock seeking new sources of water and food. Due to their nomadic lifestyle, they do not have a natural (instinctual) ability to find their way back to your home if they fly some distance away from it and will most likely perish.

14. For specifics on caring for a parakeet, see I. Birmelin and A. Wolter, *The New Parakeet Handbook* (Hauppauge, N. Y.: Barron's, 1986). Other handbooks from the same publisher specify how to care for other bird species.

15. Pepperberg provides brief examples of the social modeling procedure in most of her papers about Alex the parrot (see note 4 in Chapter 1). The most detailed description is in I. M. Pepperberg, "An Interactive Modeling Technique for Acquisition of Communication Skills: Separation of 'Labeling' and 'Requesting' in a Psittacine Subject," *Applied Psycholinguistics*, 1988, *9*, 59–76. Specifics on how to teach birds to talk meaningfully can be found in: Birmelin and Wolter, *The New Parakeet Handbook*, pp. 33–34 [14]; A. Wolter, *African Gray Parrots* (Hauppauge, N.Y.: Barron's, 1986, pp. 26–29; A. Wolter, *Cockatiels* (Hauppauge, N.Y.: Barron's, 1991), pp. 30–32; O. von Frisch, *Mynahs* (Hauppauge, N.Y.: Barron's, 1986), pp. 52–56.

INDEX

abstract concepts, formation of
  avian, 3, 9, 99
aesthetics (avian), 100
  display, 53
  musical, 46, 50, 53–54
  visual, 53–54
albatross (Laysan), migration of, 59
altruism (avian), 12–13, 44, 115–16
American Sign Language, 34, 120,
  124, 125
Animal Minds (Griffin), 163
Animal Thinking (Griffin), 161
anthropomorphism, 1, 96
  dread of, 14–17, 157, 163
ants
  aggressiveness, 140–41
  body language, 136
  and communication, 136–37
  cooperative agriculture, 141
  environment, altering, 138
  flexibility of, 139–40
  and navigation, 138–39
  odors, communication by, 136–37
  personality differences, 137–38
  scientific assumptions, 135–36,
    137, 139, 142
auditory capabilities. *See under* navigation (avian)
avian individuality
  caged pets, 107–08
  ignorance of, 104–06
  poultry, 106–07

avian/human friendships, 100
  jackdaws, 76–77
  parakeets, 78–87
  and religion, 97
  starlings, 75
  tits (blue), 88–97

Baker, R. Robin, 59, 71
beak language. *See under* communication (avian)
beavers
  and instincts, 111
bee eaters
  tools, use of, 12
Beer, Colin, 48
Berry, Thomas, 157
Bimerlin, Immanuel, 83, 84
Birds as Individuals (Howard), 96
blackbirds
  befriending, 166, 167
  communication, 166
  songs of, 95–96
  territorial flexibility, 15
  tools, use of, 12
blackbirds (European)
  songs of, 49–50
bobolinks
  and earth's magnetism, 65
body language (avian). *See under*
  communication (avian)
body language (insects)
  honeybees, 144

bowerbirds (satin)
nests, decoration of, 53–54
tools, use of, 10
buzzards (Austrian black-breasted)
tools, use of, 12

caciques
nesting flexibility, 18
caged pets, 107–08
befriending, 169–71
hazards, removing, 170
calls. *See under* musical ability
(avian)
canaries
human abilities of, 9
cetaceans
intelligence of, 129–32
Chauvin, Rémy, 137–38
Chernobyl radiation release, 155–56
chickens
domesticated, 106–07
human abilities of, 9
and imprinting, 29
chimpanzees
aggressiveness, 140
and communication, 125–29
chlorofluorocarbons
and the ozone layer, 155
Chomsky, Noam, 25–26
cichlid fish
intelligence of, 133
cleaner fish
intelligence of, 133–34
cock-of-the-rocks
human dance, influence on, 57
cognitive ethology movement, 161–62
communication (animal)
apes, 119–25
chimpanzees, 125–29
communication (avian), 150–51
beak language, 35, 99
body language, 26, 34–35, 41–42, 99
eye-talk, 34–35, 99
human/avian, 35, 39, 40–46, 80
human language, use of, 4–8, 80, 99

and mates, 44–45
research on, 39
symbolic-linguistic, 3
visual display, 35
See also musical ability
communication (cetaceans)
dolphins, 129–31
whales, 131–32
communication (human)
body language, 33–34
sign language, 34
communication (insects)
ants, 136–37
honeybees, 144–46
conceptual abilities
avian, 8–10, 99
condor (California)
and extinction, 152
Corbo, Margarete S., 75
cranes
dancing, 57
cranes (whooping)
and extinction, 152
crows
and altruism, 12
calls, distinguishing, 36
nest building, 55
protective strategies, 20
tools, use of, 11
vocalizations, 38
cuckoos
instincts of, 29–31
and migration, 72

dance language (honeybees),
145–46
dancing (avian)
human imitation of, 57
dancing (human)
Bavarian peasant Schuhplatter,
57
Native American, 57
Darwin Charles, 114, 116, 117, 142
dolphins
and communication, 129–30
intelligence of, 130–31
domesticated birds, 106–07

ducks
and imprinting, 29
ducks (Mallard)
and migration, 68
ducks (mandarin)
emotional displays, 115

eagles
tools, use of, 12
eagles (bald)
and extinction, 152
eagles (golden)
and extinction, 152
ecological destruction. *See* environment; environmental devastation
Ehrlich, Paul R., 158
Emlen, Stephen, 59
emotions and feelings
animal, 161
avian, 100, 115, 161
environment
destruction of, 152–54
exploitation of, 113–14
human attitudes toward, 157–58
ozone layer, danger to, 155
pollution, 154–55
and population, 157
radiation, 155–56
tropical rain forests, 156
environmental devastation
solutions to, 158–59
extinction (avian)
environmental poisoning, 152–54
solutions to, 158–59
eye-talk. *See under* communication (avian)

feelings and emotions. *See* emotions and feelings
finches (Galapagos Islands)
tools, use of, 12
finches (woodpecker)
tools, use of, 10–11
fish
and scientific research, 133–35
foraging adaptability, 15–16

Fox, Matthew, 158
Francis of Assisi, Saint, 97
Franklin, Dr., 115
Frisch, Karl von, 144, 147
fun. *See* play & pleasure

Gardner, Dr. Allen, 125
Gardner, Dr. Beatrix, 125
Gardner, Howard, 53
geese
calls of, 37
and imprinting, 29
geese (Canadian)
and migration, 63, 67–68
goby fish
intelligence of, 134
goldfinches
songs of, 52
Goodall, Jane, 129
gorillas
and communication, 120–25
Gould, James, 144, 145, 147
Griffin, Donald R.
cognitive ethology movement, 161, 162–63
herons bait-fishing, 12
nest repair, 18
young, defending of, 20
grouse (black)
human dance, influence on, 57
grouse (sage)
human dance, influence on, 57
gulls
foraging flexibility, 15
gulls (laughing)
calls of, 37
gulls (ring-billed)
and earth's magnetism, 65

hammerheads (African)
nest building, 55–56
Hartshorne, Charles, 48, 50
Harvard Psychological Laboratories, 8–10
hawks (broad-winged)
winter/summer home flexibility, 16

Heinrich, Bernd, 77–78, 97
Hernstein, Richard J., 8, 117
herons
  nesting flexibility, 17
  tools, use of, 11–12
herons (green-backed)
  fishing with bait, 12
Hölldobler, Bert, 136, 142
honeybees
  body language, 144
  communication, 144–46
  dance language, 145–46
  flexibility of, 146–47
  and navigation, 145, 147–48
Howard, Len
  on avian pace, 37
  on avian/human communications, 35
  avian/human friendships, 88–97
  on bird song, 47–48, 49–53, 168
  wild birds, befriending, 151–52, 165–69
Human Nature of Birds, The (Griffin), 163
human/avian friendships. *See* avian/human friendships
Huxley, Julian, 96, 105

imitation
  human, 28
imprinting, 29–31
instinct (animal), 110–12
instinct (avian), 23–24, 98–99
  imprinting, 29–31
  intelligent implication of, 31–32
  misassumptions concerning, 99–100, 110–12
  and newborns, 29–30
  self-preservation, 11
instinct (human), 23–24, 110–12
  imitation, 28
  intelligent implication of, 31–32, 112
  and language, 24–26
  and navigation ability, 69–70
  and newborns, 27–29, 30–31

pubertal program (female), 72
pubertal program (male), 71–72
intelligence (animal)
  apes & chimpanzees, 120–29
  communication, 119–29
  flexibility of, 161–62
  and philosophy, 150
  and religion, 149–50
  and research, 118–19, 162
  and science, 149–50
intelligence (avian)
  cognitive ability, 99
  conceptual, 3, 8–10, 161
  flexibility of, 3, 14–22, 100, 161–62
  humans, compared to, 3, 9, 23–24, 25, 98–99
  and philosophy, 150
  and physical size, 108–09
  and religion, 149–50
  and science, 114–17, 149–50
  small brain fallacy, 109–10
  superiorities, 101–02
  symbolic-linguistic, 3
  See also communication; life tempo; navigation
intelligence (cetaceans)
  dolphins, 129–31
  whales, 131–32
intelligence (fish), 133–35
intelligence (human)
  superiorities, 102–03
intelligence (insects)
  ants, 135–42
  honeybees, 143–48

jackdaws
  avian/human friendships, 76–77
  human abilities of, 9
  tools, use of, 11
jaegers (pomarine)
  territorial flexibility, 15
jays
  and altruism, 12
  protective strategies, 20
jays (blue)
  vocalizations, 38–39

jays (green)
    tools, use of, 11
jays (pinyon)
    mate choice, 21
jays (scrub)
    and altruism, 44
    communication, 40–46

Kastner, Joseph, 83
kingbirds (Eastern)
    winter/summer home flexibility,
        16
kingfishers (ruddy)
    nesting flexibility, 18
Kuhn, Thomas, 163

Lakatos, Imre, 163
Laudan, Larry, 163
learning set tests, 9
Leek, Julian, 76–77
Leslie, Robert, 40–42
life tempo (avian), 101–02
Lindauer, Martin, 144
linnets
    songs of, 50–52
Loeb, Jacques, 116
lovebirds
    and altruism, 12
Lovelock, James, 158

magpies
    aggressiveness of, 166
manakins
    dancing, 57
Markl, Hubert, 145
martins (house)
    nesting flexibility, 18
mating erotically (avian), 3
memory (avian), 10
migration. See navigation (avian)
mockingbirds
    and imitation, 50
Morgan, C. Lloyd, 116
multiple reversal tests, 9
musical ability (avian), 3
    aesthetics of, 46, 50, 53–54
    calls of, 26, 35–37

chattering & muttering, 38–39,
    43, 44
creative singing, 81, 100
as entertainment, 51–53
high & low frequency, 36
and human sensory limitation,
    48–49
human song, compared to, 47–48
rapid succession, 36–37
singalong, 43
songs, 26, 37–38, 46–53
subsong, 25
in unison, 52–53, 100
musical ability (whales), 131–32

narcissism (human), 112–13
Native American dance
    avian influence on, 57
navigation (avian), 3, 58–60, 101
    auditory senses, 67
    and experience, 68
    instinctual background, 71–73
    and magnetism, 64–66
    olfactory senses, 66
    overview of, 68–69
    and rational decision, 67–68
    special senses, 59
    by stars, 61–62
    sun as a compass, 60–61
    visual landmarks, 63–64
    wind & weather, 62–63
navigation (human)
    instinctual ability, 69–70
navigation (insects)
    ants, 138–39
    honeybees, 143, 147–48
nest building
    craftsmanship, 54–56
    decoration of, 53–54
    and predators, 17, 18–20
nesting (avian young)
    diets, adjusting of, 16, 100
    limiting, 21–22
    protecting, 14
    teaching, 20–21
nesting flexibility, 17–18, 100
New Parakeet Handbook, The, 83, 87

nightingales
protective strategies, 18–19
song of, 168
nuthatches (Clark)
memory of, 10
nuthatches (North American)
tools, use of, 11
nuthatches (pigmy)
nesting flexibility, 18

oddity tests, 9
olfactory capabilities. *See under* navigation (avian)
Origin of Species (Darwin), 114
ostriches
dancing, 57
ovenbirds
nest building, 55
owls (great horned)
avian/human friendships, 77–78
vocalizations, 39
owls (tawny)
offspring, limiting of, 22
ozone layer
and chlorofluorocarbons, 155

Papi, Floriano, 66
parakeets
avian/human friendships, 78–87, 169–71
communication of, 80, 85–86, 99, 171
creative singing, 81
flying with joy, 85
and masturbation, 86–88
play, 84–85
sexual activities, 81–84
parrots
communication of, 99
human abilities of, 9
interrelationships, 15–16
parrots (African gray)
communication of, 4–8
parrots (kea)
foraging flexibility, 15–16
Patterson, Francine (Penny), 120–22

Payne, Roger, 132
peacocks
display of, 53
penguins
young, teaching of, 21
Pepperberg, Irene M., 4–8, 79, 117, 171
peregrine falcons
nesting flexibility, 17
young, teaching of, 21
petrels (fish-eating)
olfactory capabilities, 66–67
pheasants (Argus)
display of, 53
pheromones (communication by odors), 136–37
philosophy
and animal intelligence, 150
pigeons
conceptual abilities of, 8–10
pigeons (homing)
auditory capabilities, 67
and earth's magnetism, 65–66
and migration, 60, 68
olfactory capabilities, 66
pigeons (tooth-billed)
nesting flexibility, 17
play & pleasure (avian), 3, 42–43, 84–85, 100
dancing, 57
flying with joy, 85
games, 56–57
and mirrors, 86–87
pollution
and avian extinction, 152–55
ozone layer, danger to, 155
radiation, 155–56
population
and the environment, 157
protective strategies, 18–20
pubertal program (human female), 72
pubertal program (human male), 71–72

Questions of Animal Awareness, The (Griffin), 161

radiation
  and the environment, 155–56
ravens
  tools, use of, 11
religion
  and animal intelligence, 149–50
  and avian/human bonds, 97
research projects
  Alex (African gray parrot), 4–8, 79, 117
  chimpanzees, 125–29
  Koko (gorilla), 120–25
  Lorenzo (jay), 40–46
  reward substitutes, 5
robins
  befriending, 167
robins (European)
  and altruism, 12
  and earth's magnetism, 65
Romanes, George, 114–15, 116, 117, 142
rooks
  tools, use of, 11
Rose of Lima, Saint, 97
ruffs
  human dance, influence on, 57
Sahtouris, Elisabet, 158
Savage-Rumbaugh, Sue, 127
science
  and animal intelligence, 149–50
  and avian individuality, 105–06
  and avian research, 114–17
  fundamental assumptions, 162
scientific names of bird species, 173–75
Seeley, Thomas, 144, 145
self-preservation instincts (avian), 11
senses (avian)
  high & low frequency, 36
  and ultraviolet light, 64
  unknown special senses, 59
  See also navigation (avian)
senses (human), 28–29
  limitation of, 48–49

sequential order, 9
sexual activities
  and eroticism, 3
  masturbation, 86–88
  parakeets, 81–84, 86–88
  sparrows, 83–84
shearwaters
  olfactory capabilities, 66–67
shearwaters (great)
  migration of, 58
shearwaters (Manx)
  migration of, 59
Sherrard, Philip, 150
shrikes (bou-bou)
  songs of, 52
skills (avian), 101
Skutch, Alexander, 20, 46
  on bird song, 48
songs. See under musical ability (avian)
sparrows
  migration of, 58–59
  nesting flexibility, 17
  sexual activities, 83–84
  songs of, 49
species (avian), scientific names of, 173–75
starlings
  avian/human friendships, 75
starlings (European)
  communication of, 99
  migration of, 68
storks
  territorial flexibility, 15
subsong. See under musical ability (avian)
sunbirds (golden-winged)
  territorial flexibility, 15
swans (black-necked)
  and extinction, 152
symbolic-linguistic intelligence
  avian, 3
tailorbirds
  nest building, 55
terns
  and altruism, 12, 115

territory (avian)
    defending, 14
    flexibility toward, 15, 16–17
tests (avian)
    learning set, 9
    multiple reversal, 9
    oddity, 9
Thorndike, Edward, 116
thrushes
    tools, use of, 12
thrushes (hermit)
    songs of, 48
titmice (penduline)
    nest building, 55
tits
    and covered milk bottles, 16
tits (blue)
    avian/human friendships, 88–97
    befriending, 167
    communications, 92
    individuality & personality, 93
    sexual activities, 91
tits (great)
    avian/human friendships, 166
    foraging flexibility, 15
    human abilities of, 9
    individuality & personality, 94
    songs of, 51
tools, use of (avian), 10–12
tropical rain forests
    destruction of, 156
turkeys
    domesticated, 106–07

U.S. Environmental Protection
    Agency, 156

visual display. See under communica-
    tion (avian)
vultures (Egyptian)
    tools, use of, 12

Waal, Frans de, 128
Walcott, Charles, 59, 69

warblers (bay-breasted)
    winter/summer home flexibility,
        17
warblers (black-throated)
    nesting flexibility, 18
warblers (Cape May)
    winter/summer home flexibility,
        16–17
warblers (Sylvia)
    and migration, 60
warblers (willow)
    song of, 168
Watson, John, 116
weaverbirds
    nest building, 55
weaverbirds (black-headed)
    nesting flexibility, 18
Western technology and environ-
    mental exploration, 113–14
whales
    musical intelligence of, 131–32
Wheeler, William Morton, 136,
    142
wild birds
    befriending, 165–69
    and science, 104–06, 114–17
    taming of, 151–52
Wilson, Edward O., 136, 137, 142
Wilson, Sheryl C.
    birds, befriending, 97, 106, 165
    parakeets, observations of, 79, 80,
        84–85, 169–71
winter and summer home flexibility,
    16–17
Wolter, Annette, 84, 87
woodpeckers
    human abilities of, 9
    nesting flexibility, 18
wrens (marsh)
    songs of, 49
wrens (winter)
    songs of, 49

Zoologist (scientific journal), 115